Synthesis Lectures on Mechanical Engineering

This series publishes short books in mechanical engineering (ME), the engineering branch that combines engineering, physics and mathematics principles with materials science to design, analyze, manufacture, and maintain mechanical systems. It involves the production and usage of heat and mechanical power for the design, production and operation of machines and tools. This series publishes within all areas of ME and follows the ASME technical division categories.

Mark Dingman

Fundamentals of Automotive Structures and Battery Electric Vehicle Applications

 Springer

Mark Dingman
Daphnis Structural Solutions
Royal Oak, MI, USA

ISSN 2573-3168   ISSN 2573-3176   (electronic)
Synthesis Lectures on Mechanical Engineering
ISBN 978-3-031-75932-1   ISBN 978-3-031-75933-8   (eBook)
https://doi.org/10.1007/978-3-031-75933-8

This Springer imprint is published by the registered company Springer Nature Switzerland AG
The registered company address is: Gewerbestrasse 11, 6330 Cham, Switzerland

If disposing of this product, please recycle the paper.

Preface

I was deeply involved in the development of BEVs when this book began formulating in my head. My job at the time was to lead the identification of effective, efficient structural solutions for BEVs and to advocate for the leadership decisions that were necessary for those solutions. It had become evident after months of reviewing computer simulation and reverse engineering competitive structural executions that a BEV is not simply an ICE with different propulsion content. The very nature of a BEV changes some fundamental engineering parameters and, as such, understanding the foundational physics of a BEV is a prerequisite for the development of a great BEV.

I witnessed manufacturers introduce well intentioned BEVs that were significantly heavier than their segment leader and more expensive to produce. These designs indicated that they were facing the same obstacles that my teammates were encountering.

As my involvement with programs continued and I dug deep into troublesome structural loadcase after troublesome loadcase, the extent to which the BEV 'changes the story' grew. I began to visualize more and more chapters to share what I had learned; to share how a BEV affects an automobile's structural strategy and how to approach structural development for success. The intended audience was automotive structural engineers working on electrified vehicles.

Things changed as I began writing the book, however. I found that there was a near never-ending trail of background information that needed to be included before I addressed the more complex topic of BEV structural topology and structural behavior. The book quickly transformed from something that only a subset of automotive engineers would find valuable to, quite honestly, the book I wish I could have read as a newly hired automotive structures engineer.

To be blunt, although for good purpose, undergraduate mechanical engineering degrees tend to be rather broad and much of the physics covered is generic. As great as my undergraduate learning was, there was a gap between it and the world of automotive structures; a gap between engineering problems involving relatively simple objects &

boundary conditions and what is seen in automotive industry loadcases. Given this, the book's content has expanded significantly since its original conception.

Cost efficiency of automobiles is critical for the long-term viability of their manufactures and understanding the fundamentals of structural efficiency is a critical element of that goal. I offer this book as tool for their vehicle development community and to improve the odds of their success.

Royal Oak, MI, USA   Mark Dingman

Acknowledgments

Acknowledgments This book captures a good chunk of what I have learned through the years as an automotive structural engineer, and I am grateful to those who taught me and supported me throughout its writing.

I must first and foremost thank my wife Mary (Fig. 3.9) for her never ending encouragement and support. There were many times where I felt overwhelmed and she knew how to keep me going. I must also extend gratitude to my daughters (Figs. 1.10 and 5.18). Their excitement was very helpful and of course, my parents (Fig. 1.1).

I am aware that I entered the automotive workforce in a rare environment and grateful for it; surrounded by structural experts like Albert Chou who taught me to look beyond the superficial results of computer simulation and to how to use simulation to learn why structure was behaving the way it was. Or Stan Serpento, who would host lunchtime nerd-out sessions where we dissected equations from old textbooks to understand vehicle behavior.

I'm also thankful for colleagues who never shied away from spending time around a computer, a post-test vehicle, or a competitive automobile; contimplating structural behavior and uncovering the keys to a successful, great design. These include Giles Bryer (Fig. 3.12), Brian Callaghan (Fig. 3.34), Nick Kriete (below), and Andre Matsumoto (below) all of which also provided feedback on this book's content.

And finally, thanks to Ronny Karlsson and Mike Regiec (also illustrated below) who leveraged their expertise and experience to provide feedback.

Introduction

As suggested in the Preface, this book covers both an introductory level of foundational automotive structure information and dives deep into the governing physics of influential loadcases and BEV peculiarities; appropriate for both new and seasoned personnel involved in automotive structural development. These topics are covered primarily from the perspective of the body-frame-integral structural configuration, however the foundational lessons of this book can be applied broadly.

Presentation of the material is focused on building the reader's structural intuition, as this is central to this book's purpose. Its lesson is to understand your medium; understand that great design flows from physics, material, & geometry and commit yourself to learning & retaining that knowledge.

Great painters understand their medium; how paint behaves on canvas and with the brush. Great teachers are deeply familiar with their subject matter and how to present that material clearly to students. Automotive structure is very similar, in my mind. A great automotive engineer understands their medium, including metal, composites, geometry, the manufacturing process. Great automotive structure designs do not happen by chance.

Learn from others, keep an open mind, and draw a FBD when all else fails.

Contents

About the Author

Mark Dingman (Fig. 3.4) started his automotive career as a computer simulation engineer at the Saturn Corporation after completing an undergraduate degree at Rensselaer Polytechnic Institute. He later joined General Motors, received a master's degree from Purdue University and spent a total of 29 years engaged in automotive structural development. Mark has held many different structural engineering positions including; managing GM's global structural strategy, leading structural solution development for global architectures, and leading development of performance remediation solutions for new loadcases.

Mark started the consulting company Daphnis Structural Solutions in 2023 and enjoys a continued, helpful presence in the automotive industry.

He's a husband, a father, an amateur artist, a native of Middleport New York, and can be easily distracted by offers to go biking, hiking, or snowboarding.

Abbreviations

ANCAP	Australian New Car Assessment Program
BEV	Battery Electric Vehicle
EA	Energy Absorption (as in, "EA-foam")
ENCAP	European New Car Assessment Program
FBD	Free Body Diagram
FCEV	Fuel Cell Electric Vehicle
FMVSS	Federal Motor Vehicle Safety Standers (US)
ICE	Internal Combustion Engine
IIHS	the Insurance Institute for Highway Safety
KE	Kinetic Energy
kWh	Measure of battery size (power delivered in one hour of time)
Latin NCAP	Latin American New Car Assessment Program
NCAP	New Car Assessment Program
NHTSA	National Highway Transpiration Safety Administration (US)
NV or N&V	Noise and Vibration
USNCAP	United States New Car Assessment Program

Automotive Structure Foundations

Abstract

This chapter introduces the reader to a broad background of automotive structure information and concepts and acts as a foundation for the subsequent loadcase chapters of this book. Readers new to the world of automotive structure will benefit from the topics 'nomenclature and terminology of automotive structure', 'automotive structure materials and how the structures are manufactured', and 'automotive structure configurations'. The chapter answers the question, "What constitutes structure?" before focusing on the concept of structural loadpaths, structural topologies, and how to construct them. Readers already possessing automotive structure experience are not left out, as the chapter culminates with an overview of a systematic process to design efficient structures.

1.1   Nomenclature

Global Directions

The industry uses a coordinate system to define directions within the vehicle. The x-axis runs longitudinally with rearward being positive, the y-axis runs laterally with the positive direction pointing to the right, and the z-axis runs vertically with upward being positive. The origin is typically low and in front of the vehicle, such that X and Z positions are always positive. The longitudinal and lateral directions are sometimes referred to as "fore-aft" and "cross-car", respectively. Axes are sometimes included in automotive images to orientate the viewer, as seen in the right side of Fig. 1.3. In these cases, the origin of

Supplementary Information The online version contains supplementary material available at https://doi.org/10.1007/978-3-031-75933-8_1.

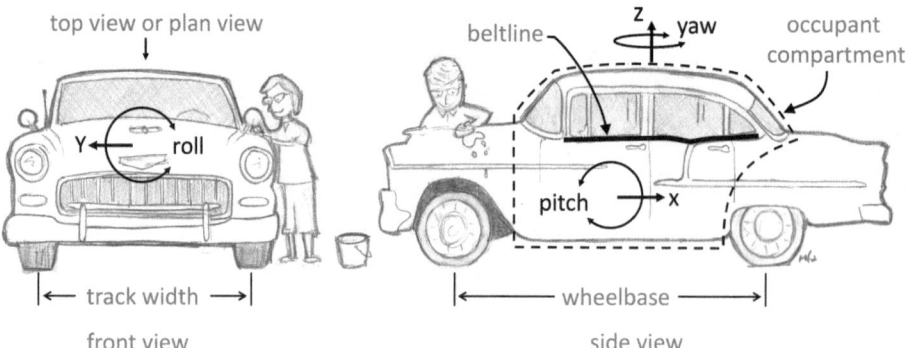

Fig. 1.1 Select vehicle attributes

the axes is not necessarily correctly positioned; the axes are included simply to orient direction. An alternative method to orient the viewer is to include an arrow and the text "front of car", as seen in Fig. 1.7.

"Pitch", "roll", and "yaw" describe rotation, about the x, y, and z axis respectively. These are illustrated in Fig. 1.1.

Vehicle Level

The term "vehicle" and "automobile" describe the complete assembly; what exits the manufacturing plant. These terms are used interchangeably in this book. Figure 1.1 illustrates three attributes of the vehicle that are used in this book; the wheelbase, track width and beltline. The wheelbase is the horizontal measurement from the front wheel center to the rear wheel center. The track width is the lateral measurement between tire centers. An automobile has a front track width and a rear track width. The distance is measured between the two front tire centers and the two rear tire centers, respectively. In many cases, a vehicle's front track width and rear track width are not the same. Also referenced in this book is the automobile's "beltline"; the location just below the side door glass.

Body Structure

The body structure is typically referred to as the "Body-In-White" (BIW) within the body engineering community. The term originates from the appearance of a body after the painting operation. Although some body structure braces and reinforcements are often installed after painting, only body structure components assembled prior to painting are included in the classification "BIW". The term "BIW plus bolt-on braces" describes the complete body structure content.

Note that the door structures are traditionally installed on the body prior to the paint operation and are considered to be part of the BIW by a manufacturing engineer; the term BIW can have different meaning depending on who is in the conversation.

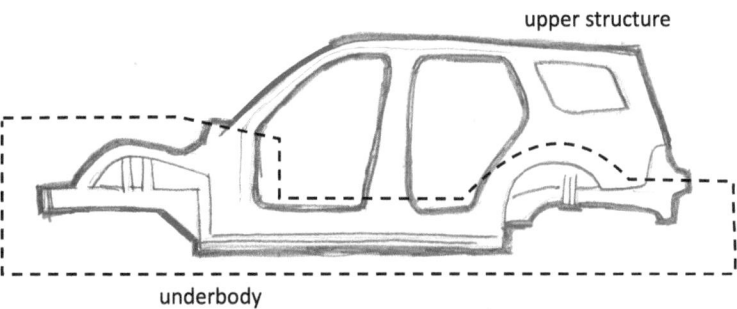

Fig. 1.2 Body structure zones

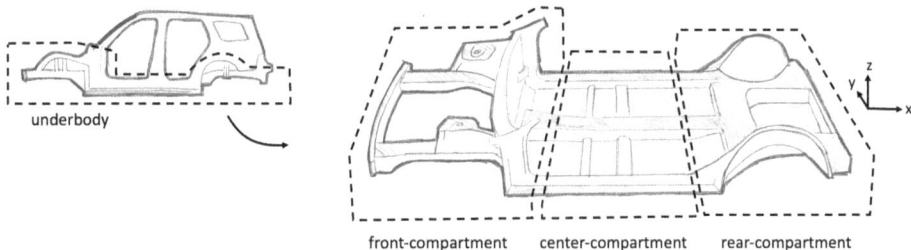

Fig. 1.3 Underbody zones

The body structure can be subdivided into two distinct zones; the underbody and the upper structure, as illustrated in Fig. 1.2. The upper structure is sometimes called the "greenhouse" or the "top-hat".

The underbody and upper structure can be further subdivided. The underbody is comprised of the front-compartment, center-compartment, and the rear-compartment, as illustrated in Fig. 1.3. Note that, technically, these names are "BIW-front-compartment", "BIW-center-compartment", and "BIW-rear-compartment", as the vehicle itself can be subdivided into zones called the "front-compartment", "center-compartment", and "rear-compartment".

The upper structure is comprised of the bodyside and roof and roof-bow components. Depending on the manufacturing strategy, the bodyside can be further subdivided into separate assemblies; the content of which can vary from vehicle to vehicle. Figure 1.4 Illustrates the condition where the bodyside has three separate sub-assemblies.

Figures 1.5 and 1.6 diagrams the names typically used to describe components of the body structure. Note that the name 'rail' is used for members that run longitudinally in vehicle and 'bar' is used for members that run laterally in vehicle.

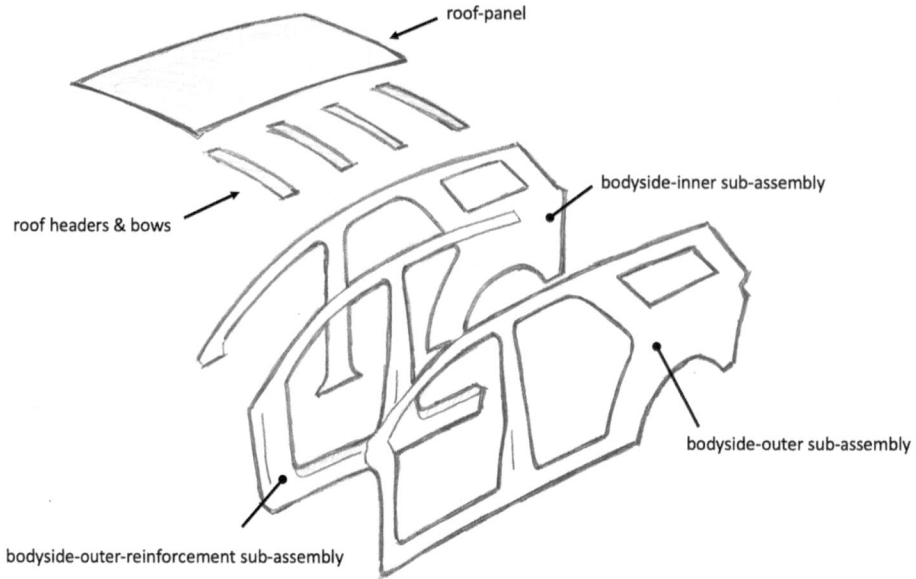

Fig. 1.4 Upper structure zones

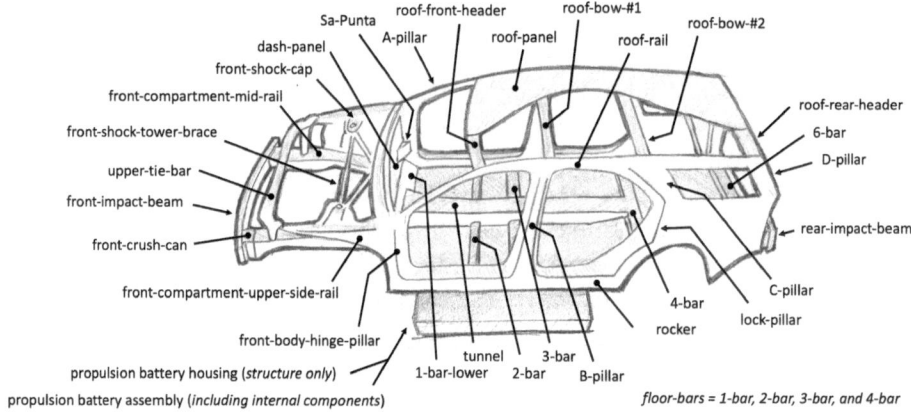

Fig. 1.5 Names of typical body structure components

Names of Body structure components are not standardized within the automotive industry. Some manufacturers use other names for the parts in Fig. 1.6. Table 1.1 captures some of the other names used for these body structure components.

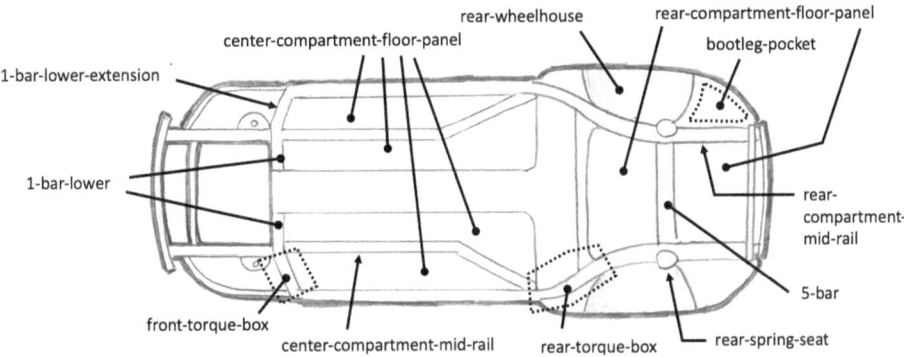

Fig. 1.6 Names of typical body structure components

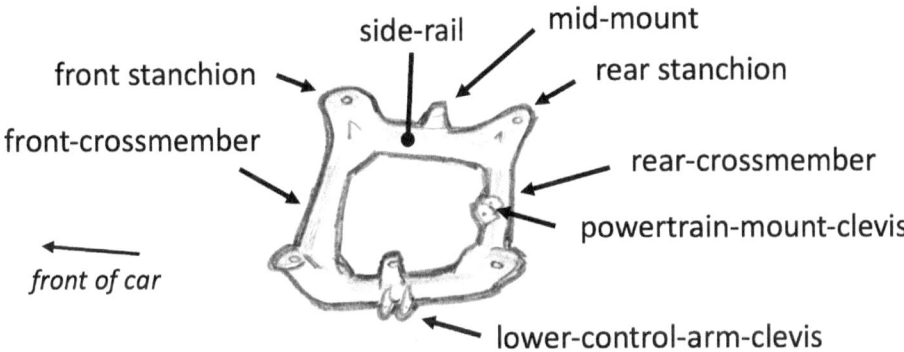

Fig. 1.7 Names of typical front cradle components

Table 1.1 Alternative names for structural components

Component name	Other names used for this component
Roof-front-header	*Windshield-header*
Roof-bow-#1	*B-bow, or 1-bow*
Roof-bow-#2	*C-bow, or 2-bow*
Roof-rear-header	*D-bow*
Roof-rail	*Cant-rail*
Rocker	*Sill*
Upper-tie-bar	*Grill opening reinforcement (GOR)*
Front-compartment-upper-side-rail	*Shot-gun*
Front-body-hinge-pillar	*A-pillar*
Propulsion-battery-housing	*Battery-box, battery-enclosure, RESS (Rechargeable Energy Storage System)*

Chassis Structure

For the scope of this book, relevant chassis structure is limited to "cradles". Cradles are structural subassemblies that attach to the body structure later in the assembly process. Traditionally, this structural subassembly has supported and somewhat enclosed the automobile's powertrain or drivetrain; thus the name "cradle".

Cradles come in two configurations, hard-mounted and isolated. Those whose interface to the body structure involves an elastomeric bushing (any type of spring element between the cradle and the body) are classified as an 'isolated cradle' whereas those that do not involve an elastomeric bushing are classified as a 'hard-mounted cradle'. Note that the presence of an elastomeric bushing is for noise and vibration performance.

Figure 1.7 defines some typical components of a front cradle. The term "stanchion" can be used to describe any formation that protrudes from the primary cradle structure to provide an attachment provision. The stanchions in this illustration provide attachment interface to the body structure.

1.2   Applied Loads

Visualize the life of an automobile. There are a multitude of conditions and requirements that the vehicle needs to manage during its life, including; maintain steering control while turning sharply, repeatedly driving over rough road without falling apart, driving over damaged concrete without producing objectionable sounds inside the cabin, and maintaining functionality after driving over a large pothole. In addition to these, some automobiles experience crashes during their lifetime which adds the requirement 'protect those inside the automobile'.

All the forementioned conditions are considered to be "vehicle level loadcases" by an automotive manufacturer. The descriptor "loadcase" is simply a condition that is considered during design and "vehicle level" describes the fact that it is a condition that occurs to the entire automobile and the applied forces are due to contact with the environment.

"Sub-system" loadcases are also considered by an automobile manufacturer during design. These are loadcases where vehicle level loads are distilled to different components of the vehicle. The sub-system loadcase 'static stiffness of the front cradle at a suspension link interface', as illustrated in Fig. 1.8, is an example of such a loadcase.

$$K_{static} = \frac{Force}{displacement}$$

where K_{static} = static stiffness.

The force applied to the front cradle occurs due to its interface with a suspension link, not the outside world. Stiffness at a suspension link interface impacts many different vehicle-level loadcases including maintaining steering control while turning sharply and driving over damaged concrete without producing objectionable sounds inside the cabin.

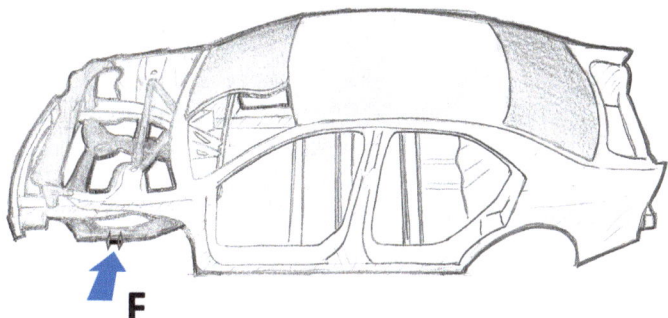

Fig. 1.8 Static stiffness of the front cradle at a suspension link interface

Sub-system loadcases have the advantage of being able to be assessed quickly on smaller, simpler computer simulation models or physical test properties.

Once engineers determine the relationship between performance in sub-system load-cases and vehicle-level loadcases, performance in the sub-system loadcases can be considered indicative of the automobile's performance in vehicle-level loadcases and sub-system performance targets can be generated.

Every automobile manufacturer considers a set of conditions, events, or "loadcases" as a part of designing their product. Many of these loadcases, in fact, are considered commonly across manufacturers. These common loadcases include those whose performance is regulated by governments; all manufacturers must meet a minimum performance level in order to sell their automobile in that country.

Loadcase Categories

The loadcases covered in this book can be grouped into categories of 'collision' and 'non-collision' events. Collision events are loadcases in which a vehicle, or vehicles, have an initial velocity and significant plastic deformation to the test vehicle is a potential outcome. We'll find in Sect. 3.2 that understanding and modeling these events is based on the Conservation of Momentum and Conservation of Energy laws.

The non-collision loadcases in this book do not involve an initial velocity. Those covered in Chap. 6 fall into one of the following categories:

- Static: a force is applied to the structure and an elastic displacement is measured.
- Quasi-Static: an applied force deforms the structure at a very low rate of speed. A 'prescribed displacement' loadcase, such as Roof Crush (Sect. 5.5), is a good example; where force is applied via a 'platen' which pushes on the vehicle at a rate of 5 mm/s until it reaches 127 mm of travel.
- Dynamic: these loadcases assess performance in the frequency range; what is the structure's first natural frequency, for example.

1.3   What Constitutes Automotive Structure

Automobiles are a complex product. There are many components in an automobile; how do we know what constitutes the structure? To make the problem more difficult, there are two categories of structure that we're interested in; global and local. Global structure is the vehicle's content that is necessary to satisfy vehicle-level requirements, like 'maintain steering control while turning sharply'. Local structure functions to satisfy local requirements and vehicle-level performance is largely insensitive to its presence. Consider a small bracket welded to the body structure to provide attachment for an electrical module, as illustrated in Fig. 1.9. This bracket does not add meaningful stiffness to the body structure with respect to vehicle-level loadcases, however it has been designed to be stiff enough such that the electrical module does not vibrate when the automobile is in operation. In this case, the bracket and the immediate surrounding body structure is considered "local structure" for this vibration loadcase.

To solidify this concept, consider the analogy of a serving tray carried by a restaurant waiter as illustrated in Fig. 1.10. We'll even consider the tray and everything on it a 'vehicle', as their function is to transport something. On the tray is a piece of cloth, two plates, two glasses of water, and two sets of chopsticks. There are a lot of requirements implied here; the tray must look presentable (thus the piece of cloth), the water must be contained (thus the glass), the tray must be stiff and strong enough to carry the items, etc. Like an automobile's requirement to 'repeatedly driving over rough road without falling apart', the requirement that the tray be stiff and strong enough to carry the items is a vehicle-level requirement. The content required to achieve that requirement is simply the tray. Here, the tray would be the 'global structure'.

Consider the plate and glass, however. These things are structure; they have strength and stiffness that provide a function in the vehicle. Consider the function and requirements that these structures provide; contain the water and contain the food. These two requirements are not vehicle level; they are specific to the particular 'local' items. Here, the plates and glasses would be 'local structure'.

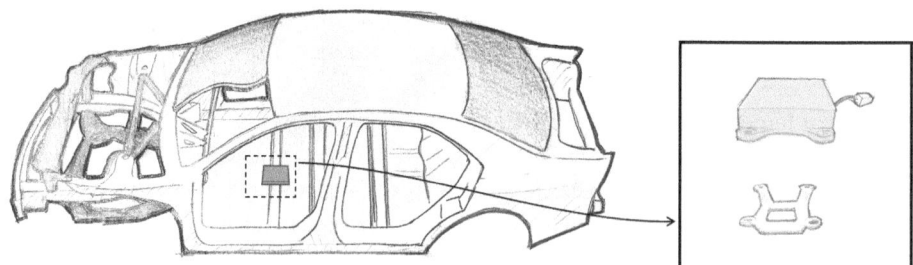

Fig. 1.9 Example of local structure

Fig. 1.10 Serving tray, a global and local structure analogy

With the complexity of an automobile and the quantity of vehicle level requirements, the definition of global structure can differ slightly between loadcases. Those uncommon conditions aside, automotive vehicle structure is generally regarded as those shown in Fig. 1.11 and listed here:

- the body structure,
- any glass bonded to the body structure,
- any braces bolted onto the body structure,
- the front and rear impact beams,
- the upper-tie-bar (sometimes bolted on),
- the instrument-panel-beam, and
- hard-mounted cradles.

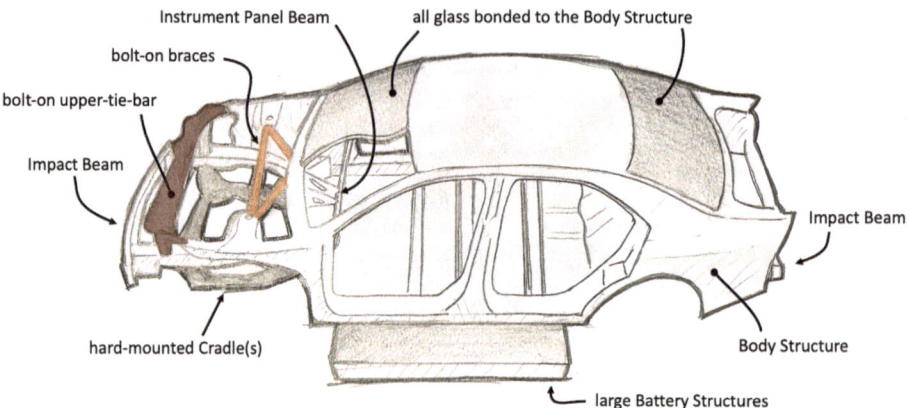

Fig. 1.11   Typical global vehicle structure

1.4   Loadpaths

The significantly loaded portions of the structure in any given loadcase defines the load-path; the loadpath is the part of the structure is being 'worked'. The loadpath is often defined graphically with a bunch of arrows drawn over an image of a structure. Loadpaths are usually specific to a single loadcase, as shown in Fig. 1.12, although some loadpath illustrations will compile the loadpaths for a few loadcases, as shown in Fig. 1.13.

Loadpath illustrations are helpful to those developing the structural design; partic-ularly as a communication and visualization tool. Here, the loadpath illustrations help those involved understand where to focus their efforts; where changes might improve

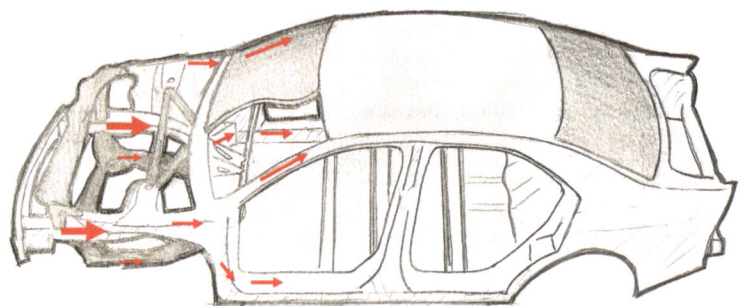

Fig. 1.12   Illustration of the loadpath for the 'Full Frontal' front crashworthiness loadcase

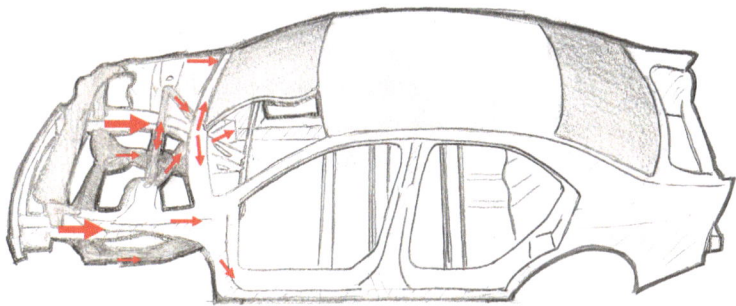

Fig. 1.13 Illustration of major loadpaths in the front compartment

performance most efficiently. These illustrations also function as a way to communicate what areas of the structure are particularly important to the wider design development community, particularly those 'packaging' the vehicle (determining the spatial allowance for every part of the vehicle, including the structure).

The free body diagram (FBD) representing a loadcase is key to identifying the loadpath. Recognizing the applied forces and constraints involved in the loadcase is critical, as the loadpath is simply the structure that lays between these two.

Although refinement of the loadpath, particularly the relative contribution of each of its elements, can be gained from commuter simulation, test data, and experience with similar structures, the origins of a correct loadpath start with the FBD.

Let's explore a few different scenarios to solidify this point.

1.4.1 Loadpath Example, Roof Crush

Roof Strength is a loadcase in which a force is applied to the *roof-side-rail* while the vehicle *rockers* are constrained in a test bed. The general test configuration can be seen in Fig. 1.14.

The orientation of the test fixture is such that there is significant vertical and lateral loading. The FBD of this loadcase represents the load directions and the boundary condition applied to the rocker, as illustrated in Fig. 1.15.

Examining the FBD, we first can identify the structure between the vertical load $F_z(t)$ and the constraint; the body *roof-side-rail, A-pillar, B-pillar, lock-pillar*, and the *rocker* (Fig. 1.17a).

Next, we can recognize that the lateral component $F_y(t)$ will have a shearing effect on the vehicle structure, encouraging the vehicle to 'match-box' or deform in the shape of the dashed lines of Fig. 1.16. By identifying the shearing effect, we can conclude that following elements of the body structure will be influential to performance; *roof-bow(s), roof-panel, front-header*, the far-side structure, and the joint stiffness/ strength between

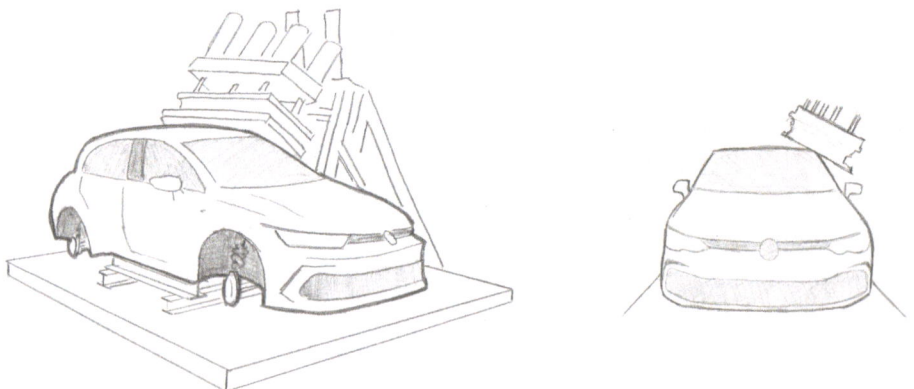

Fig. 1.14 General roof crush loadcase condition

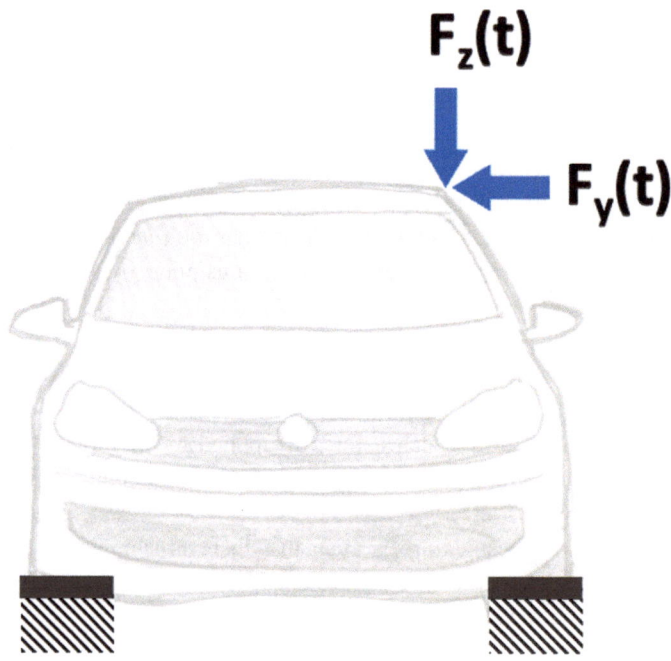

Fig. 1.15 Roof Crush Loadcase FBD

these elements. We can also identify the windshield as a structural shear surface that will contribute to performance. The major loadpath elements associated with the lateral force component are shown in Fig. 1.17b.

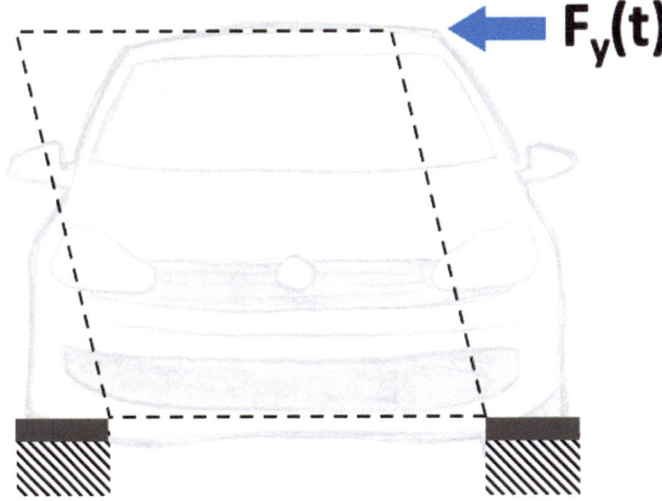

Fig. 1.16 Shear effect of the lateral force component

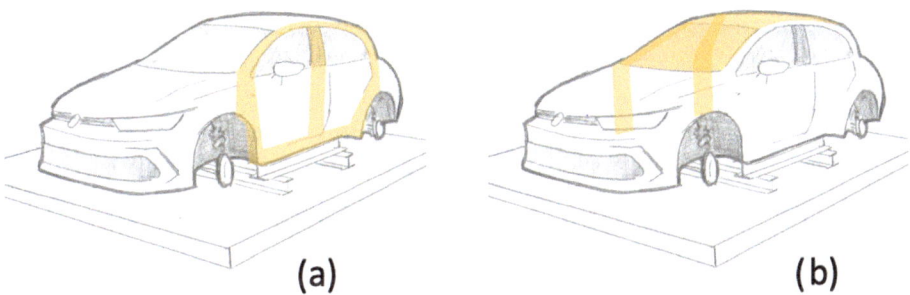

(a)   (b)

Fig. 1.17 Roof strength loadpath components

The complete loadpath for the Roof Crush loadcase is simply the compilation of the structure identified for each loading direction. Representation of this loadpath is shown in Fig. 1.18. As suggested earlier in this section, the scope of this loadpath will be tempered by insight from computer simulation, however the loadpath basis begins with an assessment of the FBD.

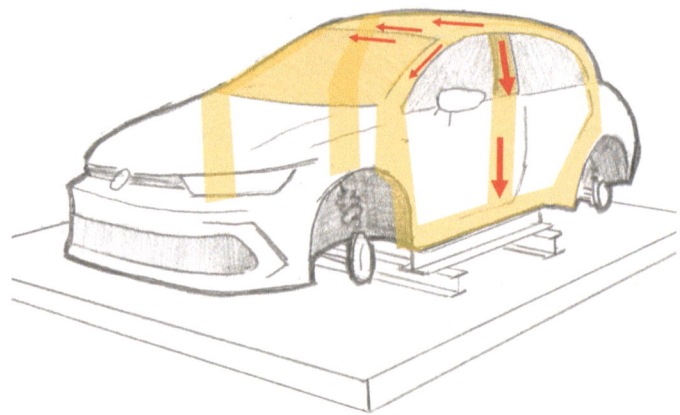

Fig. 1.18 Loadpath illustration for the roof crush loadcase

1.4.2   Loadpath Example, Full-Frontal Barrier

The method of identifying structure between applied forces and boundary conditions can also be used for loadcases for which those elements of the FBD are not immediately obvious. Consider the front crashworthiness loadcase, "Full-Frontal" in which a test vehicle collides with a stationary wall, as illustrated in Fig. 1.19.

There are two aspects of crashworthiness loadcases that must be considered when constructing a loadpath illustration, the energy absorption strategy and the energy ride-down.

Energy Absorption Strategy

There is a significant change in the test vehicle's kinetic energy in most crashworthiness loadcases; in the Full-Frontal loadcase, the vehicle has kinetic energy of $\frac{1}{2}m_{vehicle}v^2$ prior to the collision and none at the end of the event. Energy is dissipated through material fracture and crush and there are elements within the vehicle structure which have been designed to perform this function. The loadpath illustration includes these components.

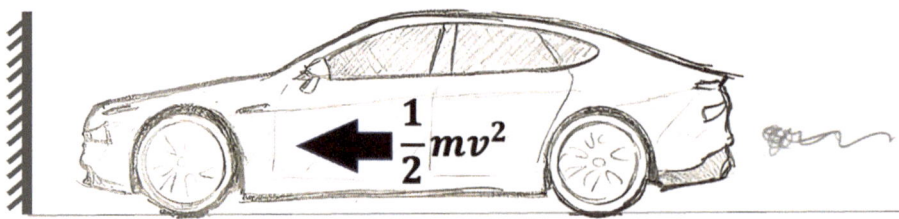

Fig. 1.19 General full-frontal loadcase condition

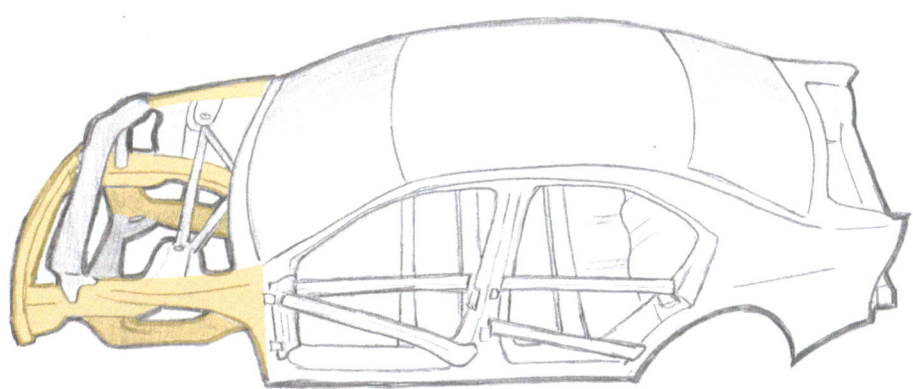

Fig. 1.20 Energy absorption structural elements for the full-frontal loadcase

The energy absorption structural elements responsible for energy absorption in the Full-Frontal loadcase are illustrated in Fig. 1.20.

Energy Ride-Down

To understand energy ride-down aspect of a crashworthiness loadcase, it is helpful to visualize the loadcase FBD at a more granular level. Recognize that what is happening in such an event is that individual masses within the automobile are being slowed down; each with their own kinetic energy of $\frac{1}{2} m_{component} v^2$. Figure 1.21 illustrates a few such masses, which can mounted components, zones of the vehicle, or occupants.

Deceleration of these individual masses apply a force on the vehicle structure per Newton's second law; $F = ma$. Thus, a FBD can be constructed representing the forces applied to the vehicle structure during the event. Figure 1.22 illustrates the forces applied by three different masses.

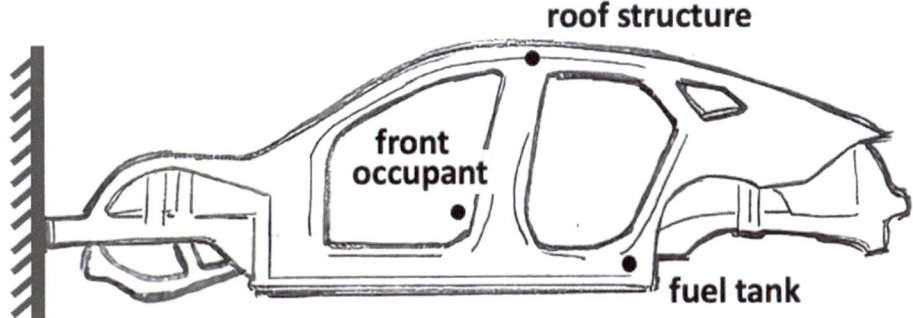

Fig. 1.21 Select significant masses

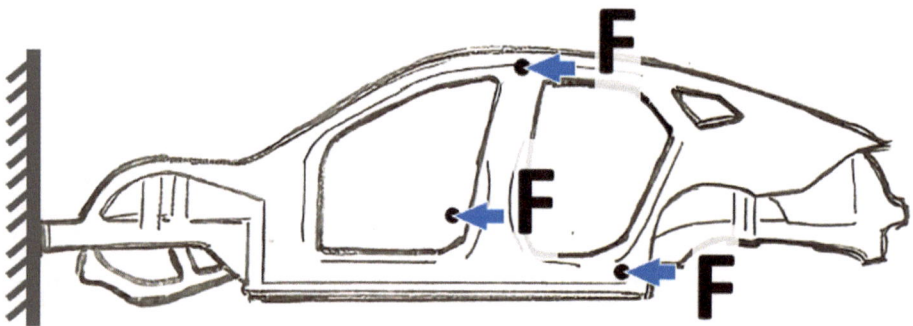

Fig. 1.22 Full-frontal loadcase energy ride-down FBD

As in the Roof Crush example, a loadpath can be constructed by drawing a line within the structure between the forces and the boundary condition. Addressing each of the forces shown in Fig. 1.22 independently.

Front Occupant

- *front-compartment-mid-rails* to *1-bar-ext* to *rocker* to *floor-bars*
- *front-compartment-mid-rails* to *1-bar* to *tunnel* to *floor-bars*
- front-cradle-side-rails to *1 bar-lower* to *1-bar-extension* to *rocker* to *floor-bars*
- *front-cradle-side-rails* to *1 bar-lower* to *tunnel* to *rocker* to *floor-bars*.

Roof Structure

- *front-compartment-upper-rails* to *front-body-hinge-pillar* to *A-pillar*
- *front-compartment-mid-rails* to *front-body-hinge-pillar* to *A-pillar.*

Fuel Tank

- *front-compartment-mid-rails* to *1-bar-ext* to *rocker* to *4-bar*
- *front-compartment-mid-rails* to *1-bar* to *tunnel* to *4-bar*

Illustrating the Loadpath

A comprehensive loadpath illustration for the Full-Frontal loadcase includes the energy absorption structural elements, as illustrated in Fig. 1.20, and those constructed above to the energy ride-down function of the vehicle structure. Figure 1.23 illustrates this composite representation.

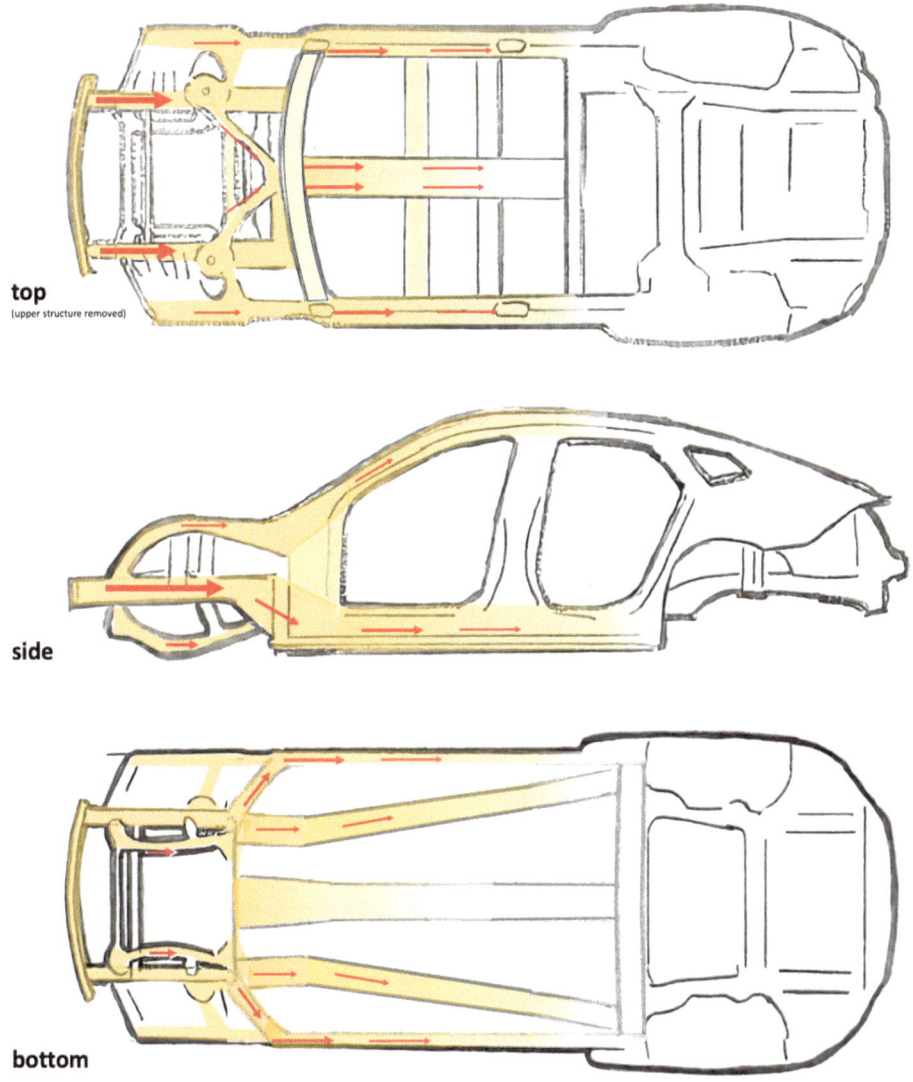

top
(upper structure removed)

side

bottom

Fig. 1.23 Loadpath illustration for the full-frontal crashworthiness loadcase

The loadpath illustration of Fig. 1.23 is shown in three views rather than one isometric view. This can be done to improve clarity. This is especially relevant in cases where the loadpath is extensive or simply cannot be represented in one view.

Things to Consider When Constructing Loadpath illustrations for Crashworthiness Loadcases.

- It's important to consider only elements of significant mass when constructing a ride-down loadpath. The entire vehicle structure would be highlighted if every element were considered and the resulting loadpath illustration would be useless for communicating critical elements of the vehicle structure.
- Consider the serving tray example illustrated in Fig. 1.10 with respect to the concept of significant masses and loadpath illustration. Visualize the contents of the serving tray and their mass; the bowls and cups are significant to a loadpath study while the wooden chopsticks can be ignored. The loadpath important for the chopsticks is captured by the heavier bowl and plate and ignoring the chopsticks does not change the result.
- Computer simulation, test results, and experience will provide guidance on what threshold is appropriate.
- It is customary to deemphasize the loadpaths further away from the barrier. This is because the amount of vehicle mass reduces as the reference moves further away from the barrier; less structure is needed to support the ride-down. Consider the analogy of a tree that must support each component of the vehicle (the loadpath illustration turned 90°), as illustrated in Fig. 1.24. As we look higher in the tree (or further rearward in the vehicle), the tree trunk diameter needed to support the components above gets smaller.
- As crush progresses, some elements will directly contact the barrier and "ground-out". They are no longer applying a forward force on the structure once this happens and do not need to be considered in the loadpath. The front suspension is an example of this in the Full-Frontal loadcase, as the front wheels ground out on the barrier during in the event. The ride-down of the front wheels' energy. occurs directly on the barrier, not through the structure. This behavior can be seen in the video provided in Link 1.1.

	Full-Frontal test of a 2020 Volvo XC60, slow motion NHTSA, USNCAP	(Link-1.1)

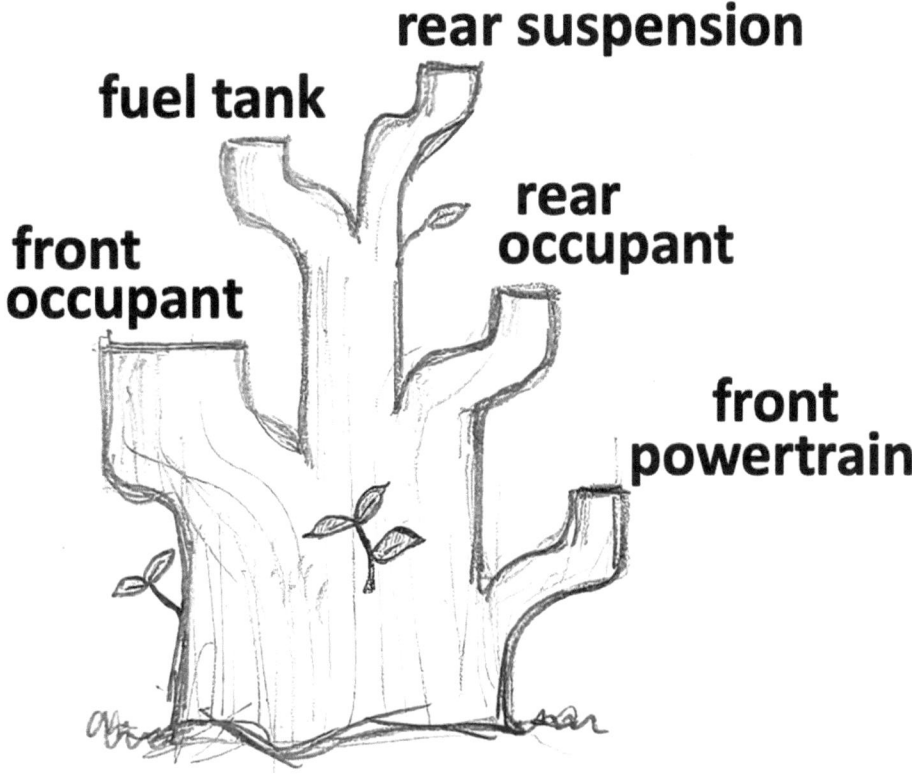

rear suspension

fuel tank

front occupant

rear occupant

front powertrain

Fig. 1.24 Loadpath girth analogy

1.5   Loadpath Topology

Loadpath illustrations are a great way to communicate critical elements of a structure and are very often found in manufacturer industry presentations, where they describe the technical aspects of a new vehicle. The depth of content is perfect for this kind of communication but the loadpath illustration very often lacks the detail needed to support the design development activity. Adding other critical, design-guiding information to the load-path illustration resolves this issue; here, we'll call this document a Loadpath Topology Strategy.

Additional information could include;

- dentification of critical sections or joints; perhaps even specifying minimum section size, strength, or stiffness
- elements of crashworthiness strategies; crush space targets, component orientation or placement, etc.

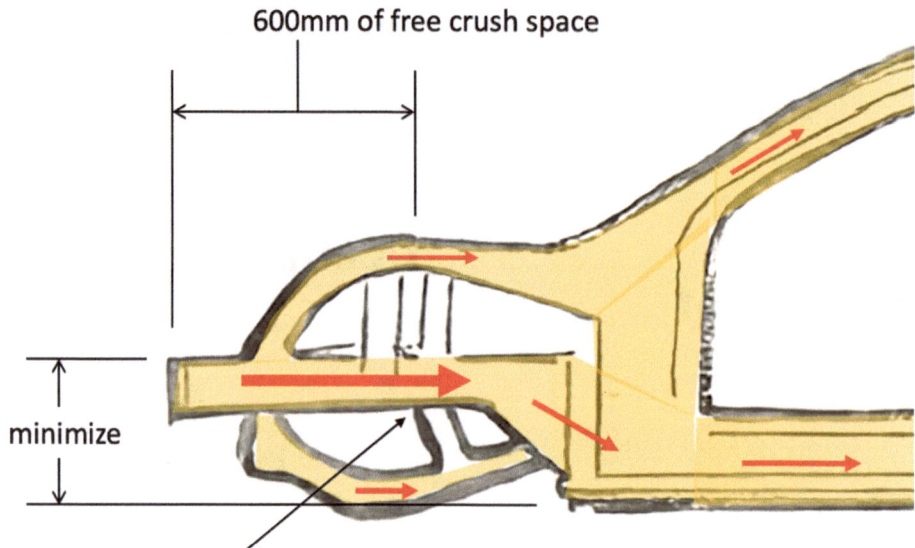

600mm of free crush space

minimize

tune attachment strength to release in Full Frontal Loadcase

Fig. 1.25 Loadpath topology strategy example

- comments about the relationship between design attributes and mass efficiency.

It's appropriate to note that there are many more contributors to a loadpath topology strategy document, including, CAE simulation studies, manufacturer design best practices, and defined manufacturer strategies for any performance area.

A simple example of loadpath topology strategy content can be seen in Fig. 1.25.

Note that this practice is not nearly as common across manufacturers as loadpath illustrations. Different manufacturers will use different tools and methods to communicate structural requirements to the rest of the design development community.

1.6   Automotive Structural Configurations

'Structural Configuration' describes the way the global structure is divided into sub-assemblies (or not). Each configuration type has advantages and disadvantages, so you will find a variety of configuration types employed across the industry.

Types of Structural Configuration

Body-on-Frame (BOF): Body-on-Frame is a configuration where the occupant compartment structure is separate from the longitudinal structure; the longitudinal structure typically being referred to as 'chassis frame structure' and the occupant compartment as

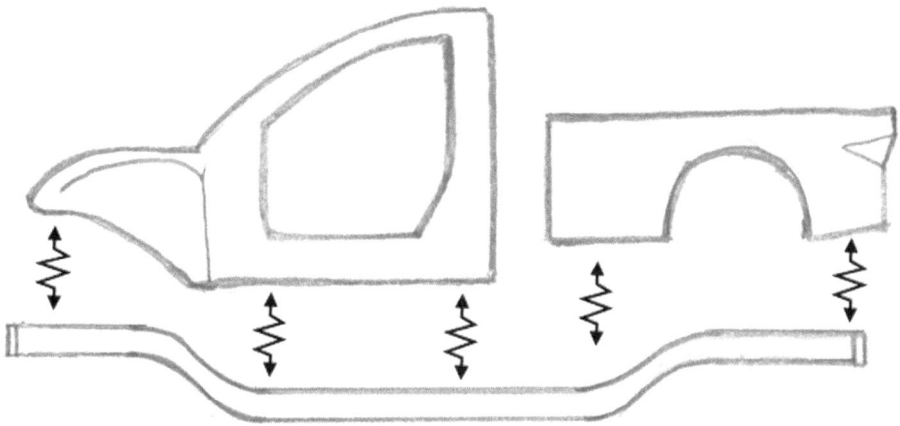

Fig. 1.26 BOF configuration

'body structure'. The two structures are traditionally connected through an elastomeric mount, as illustrated in Fig. 1.26. These mounts provide benefits such as vibration isolation but disallow the two structures to work as a unified structure. The BOF configuration is regarded as a good option for vehicles that will experience high usage loads; towing and hauling, for example.

In some instances, the body is bolted directly to the longitudinal structure without any elastomer. This condition can be qualified as a 'hard-mounted BOF' configuration.

Spaceframe: By definition, a spaceframe is a three-dimensional framework whose elements act as an integral structure. Compared to an isolated BOF configuration, a spaceframe allows the longitudinal frame structure and the occupant compartment structure to work as a singular structure. The unity between these two large structures eliminates some structural redundancy and thus enables higher mass efficiency. The Tube-Frame and 'Body-Frame Integral' configurations qualify as spaceframes.

Tube-Frame
A tube-frame is a configuration in which the entire, or nearly entire, structure is created by tubes; its structural loadpath driven by the principle of triangulation, as illustrated in Fig. 1.27a. The economics of a tube-frame manufacturing process is not conducive to large volume production. Vehicles with this configuration are rarely vehicles available as a retail product. This construction is popular in low-volume, low-budget race car production.

Body-Frame-Integral (BFI)
The components of a BFI spaceframe, illustrated in Fig. 1.27b, are more complex than the tubes of a tube-frame; typically stampings, extrusions, or castings. Joining of these components is performed using methods such as resistance spot welding and/or rivets.

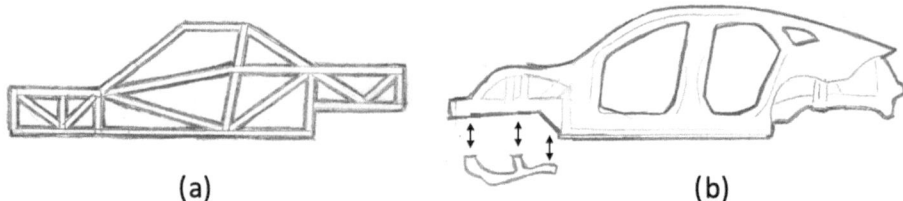

Fig. 1.27 Tube-frame and BFI configurations

BFIs are assembled using a manufacturing modularity concept where different areas of the structure are independently assembled and joining of these large sub-assemblies completes the manufacturing process. Speed is a large focus of a BFI's manufacturing process, joining processes, and component production processes, thus making the BFI configuration popular for high volume production. The BFI configuration is sometimes referred to as "unibody".

Monocoque: A monocoque structure is one where the mechanical properties the automobile's outer surfaces are significantly leveraged as structure. Vehicles using this construction typically involve a monocoque center-compartment 'tub' and metal front and rear subframes, as illustrated in Fig. 1.28. The monocoque tub is constructed of composite material. The economics of this configuration's material strategy and manufacturing processes relegate its application to low-volume, high retail price automobiles.

Subframe Qualifier: A subframe is a structural member that is constructed independently and has components which are mounted to it prior to assembling it to the rest of the structure. Automobiles constructed using any of the above structural configurations nearly always involve the use of a structural subframe. The frame of a BOF can be considered a subframe, as the powertrain, suspension, etc. are assembled to the frame prior to its 'marriage' to the body. A BFI's front cradle follows the same logic; powertrain and suspension components are mounted to it prior to the cradle being joined to the body.

Fig. 1.28 Monocoque configuration

Fig. 1.29 Stressed component

Stressed Component Qualifier: In a stressed component configuration, some vehicle component(s) also provides a meaningful level of global structure. This configuration has been historically employed in racing cars by incorporating the engine structure, or engine block, into the vehicle's global structure. This condition can be considered as a 'stressed engine subframe' and is illustrated in Fig. 1.29. Some manufacturers are exploring ways to leverage the stiffness and strength of propulsion battery cells and are designing their propulsion battery system to be a significant contributor to global structure. We can use the stressed component qualifier in these cases and call the configuration a 'stressed battery cell BFI', such as with the 2022 Telsa Model-Y.

Skateboard: The term "skateboard" has become popular to describe automotive structure however it has not been used consistently. It has been used in some instances to describe the lower portion of a hard-mounted BOF configuration and to describe the underbody portion of a BFI in others. As such, the term skateboard is more of a marketing term than an engineering one.

What is the Right Configuration?
The right configuration for any specific automobile design will depend on many factors, the most significant being;

- **production volume** (how many vehicles will be manufactured),
- **intended retail price,**
- the i**mportance of low mass to the vehicle's mission,** and
- if the vehicle **requirements include high usage loads**.

High volume vehicles demand a certain level of manufacturing efficiency in order to be profitable, here, BFI or BOF are the typical choice. Structural components can be constructed from tooling that produces parts quickly. These components can be assembled quickly, and the resulting assemblies can integrate easily into an assembly line manufacturing strategy.

Low volume production can accommodate more elaborate configurations that involve longer component production times; composite monocoques, for example. However, the associated increase in material cost and the economics of low volume production dictate that the retail price increase accordingly.

Cost efficiency is required when the vehicle's intended retail price is low; suggesting BFI or BOF construction. Alternatively, when the intended retail price is high, the 'vehicle can afford' to have more expensive structural materials and a less efficient manufacturing strategy; composite subframe configurations are an option, here.

Vehicles that expect the structure to see high loads or repetitive loads during the automobile's use tend toward the BOF configuration. The material thickness, joining type, and topology of a BOF's longitudinal rail structure has a durability performance advantage over other configurations.

1.7   Manufacturing and Materials of Automotive Structure

1.7.1   Automotive Structure Manufacturing Process

The manufacturing process used for automobile production is dependent on its sales volume expectation, region of manufacture, and labor rate. The manufacturing process for an automobile involves three high-level functions; the Body Shop, the Paint Shop, and Final Assembly. These functional processes are essentially the same regardless of annual volume (as high as 300,000 units per year or as low as 10,000 or less units per year) and region of manufacture. Multiple stations of specialized operations occur in each of these areas. Throughout the process, the vehicles being assembled travel along a "main-line" until assembly is completed and a vehicle is produced. The required material, parts, and sub-assemblies are delivered to the facility as close as possible to the point of use. It is at these locations where parts or components are loaded and fastened to the evolving vehicle. Figure 1.30 illustrates the layout of a typical automobile manufacturing facility. Whether high volume or low volume and high labor rate or low labor rate, these three manufacturing functions remain the same; just the processes within are scaled to match the production rates.

The following video links provide examples of such processes and facilities.

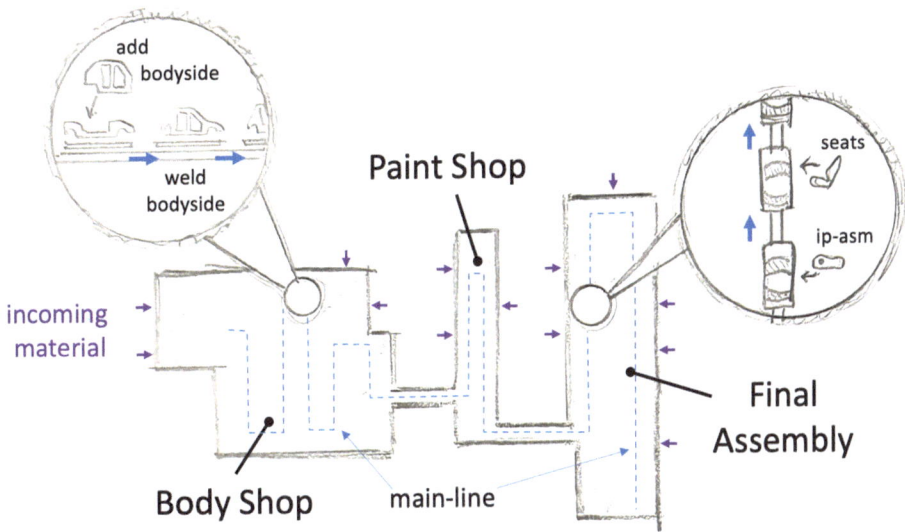

Fig. 1.30 Layout of a typical high volume manufacturing facility

	Automobile Manufacturing Plant Tour, 6 min long (https://www.youtube.com/watch?v=Zn6scKf7k_0) Engineering World YouTube channel	(Link-1.2)
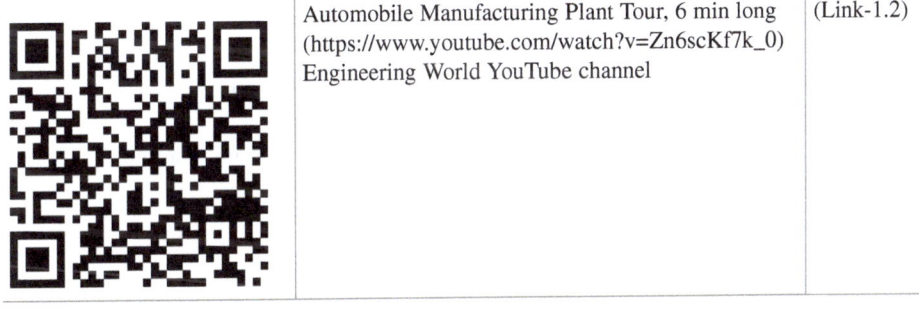	Automobile Manufacturing Plant Tour, 48 min long (https://www.youtube.com/watch?v=F3FUH9fVC_M) Free Documentary YouTube channel	(Link-1.3)

Body Shop

The Body Shop's function is to assemble the Body-in-White (BIW) structure and install closure structures. Material enters the Body Shop in the form of stampings or sub-assemblies typically produced offsite. In some cases, BIW structure sub-assemblies or assemblies are also produced in the Body Shop, a decision typically driven by logistics. The process of assembling the complete body and closure structures tends to be highly automated in high volume and high labor rate facilities. Robots are used to perform material handling, sealing and adhesive dispense, welding, and other joining processes. The automation rate is reduced when manufacturing in low volume or lower labor rate facilities. In either scenario, the Body Shop's finished product is a completed body structure with installed closure assemblies, ready for the Paint Shop.

Paint Shop

Paint Shop is the location where the BIW is cleaned, its surfaces prepped, and primer ("e-coat"), sealer and paint are applied to the body structure and closures. The process typically involves:

- Cleaning: Phosphate spray surface preparation.
- E-coat: Electrostatically applying a corrosion prevention coating to the body structure and closures.
- Sealing: Applying a relatively thick bead or ribbon of sealer to joints within the body structure and closures to protect against water intrusion and corrosion followed by
- Baking: Curing the sealer by heating the BIW and closures to a certain temperature for a certain amount of time.
- Topcoat and clear coat paint: spray application of the appropriate material
- Baking: Curing the paint through a second baking operation.

Final Assembly

All the components necessary to complete the vehicle are applied in Final Assembly; the carpeting, the instrument panel assembly, the seats, the powertrain, the chassis structure, etc. Final Assembly is sometimes referred to as "General Assembly".

Very often, components are installed as sub-assemblies or 'modules', arriving to the mainline in the sequence required for the queue of vehicles in process. This strategy minimizes the length of the assembly main-line process.

The front chassis module is a good example of the module concept; illustrated in Fig. 1.31. Here, the front cradle acts as a foundation for the sub-assembly; the front suspension and powertrain components are assembled to the front cradle prior to its arrival to the main-line. The module is installed to the body structure as a single unit.

The manufacturer of high-volume automobiles has followed this general approach for decades. However, a resurgent interest in minimizing cost has prompted many manufacturers to explore modified or dramatically different manufacturing processes, such as,

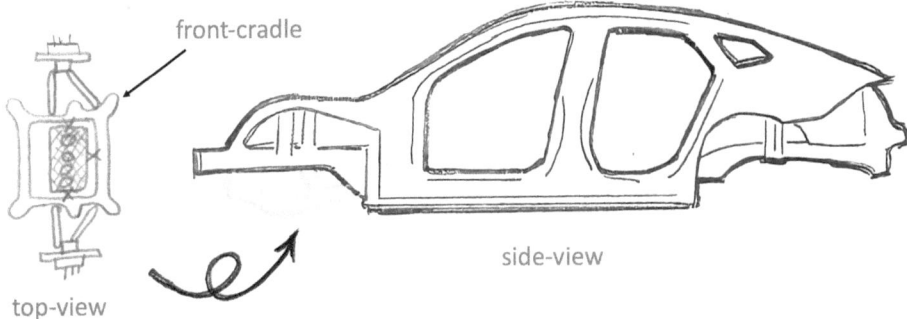

front-cradle

side-view

top-view

Fig. 1.31 Marriage of the front chassis module to the body structure

combining body shop parts into lightweight castings or redefining the traditional module boundaries. These manufacturing alternative are intended to shorten the main-line processes and reduce the time required to prepare the facility to produce a different automobile. Time will tell if this traditional high-volume manufacturing process described above will continue to be 'typical'.

Low-Volume Automobiles

The forementioned high volume process which is typically optimized for an annual production rate on the order of 250,000 vehicles per year, is not economically or logistically rational for manufacturing of automobiles produced in low quantities such as the 2022 Rimac Nirvana produced at a rate of 1500 vehicles per year. Imagine a plant designed to produce the Toyota Camry being dedicated to Nirvana production; each Nirvana would sit idle in each assembly station for about six hours after the station's work had been complete. In short, a low-volume automobiles do not require a large manufacturing footprint.

Production of low-volume automobiles typically occupy a small portion of a manufacturing plant which is also producing different vehicles or is built in a small facility dedicated to its production.

Construction of a low-volume body structure is also likely to be less automated. Robots have a high initial investment cost and are best rationalized economically if they are in use for large periods of time or in regions of manufacture with high labor rates. Low-volume body structure production would likely involve significant robot idle time; thus, these structures are produced with reduced automation levels in regions of high labor rates or predominantly with manual labor in regions of low labor rates.

Final Assembly of low-volume automobiles will occur in fewer stations, where more components and sub-assemblies are installed at each station. For example, The Rimac Nirvana's Final Assembly facility in Zafreb Croatia consists of approximately eight stations, where the automobile spends significantly more time in each station than in a high-production GA facility.

It should be noted that the use of manual labor is not strictly a function of economics. The marketing value of a product being "hand built" can be significant enough that manual labor is used in regions of high labor costs, particularly for high priced, ultra-low volume vehicles.

1.7.2 Automobile Structure Materials

Body Structure

There are a variety of different material and construction strategies that are applied to automotive structure. A typical steel automotive body structure is comprised of many individually manufactured stamped parts that are joined together in the Body Shop by welds. A typical aluminum body structure can contain stampings, extrusions, and castings which are joined together by welds, rivets, screws and adhesive. Body structures sometimes utilize both steel and aluminum in their design and can be referred to as a 'mixed-metal' material strategy, as illustrated in Fig. 1.32a. They can also involve metal and composites, whereas they could be referred to as a 'mixed-material' solution, as illustrated in Fig. 1.32b.

The appropriate material and construction strategy is largely dependent on the performance expectations of the automobile, its production volume, and its intended retail price.

The body structure of high-volume automobiles with a low retail price is traditionally constructed of steel because the material is relatively inexpensive and the cost of a large, highly automatized Body Shop can be amortized over the entire product volume. Lighter materials are introduced into the body structure as the importance of things such as driving dynamics and fuel efficiency/BEV range efficiency increases.

The material and construction strategies for low-volume automotive body structures are typically borne from the desire to minimize Body Shop tooling costs. The specific strategy taken for each low-volume automobile will vary depending on things such as; the automobile's performance expectations, capabilities of an existing manufacturing facility, and the manufacturer's experience with particular materials. Low-volume automobiles

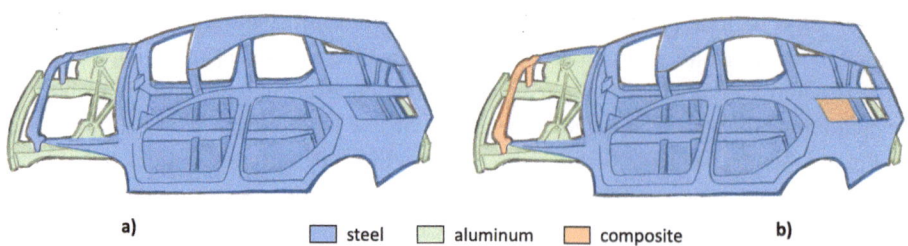

a)   ☐ steel   ☐ aluminum   ☐ composite   b)

Fig. 1.32 Example material strategies

also tend to have a high retail price which can offset the use of more expensive body structure materials such as composites.

Cradle Structure

The rationale for material and construction strategy for cradles is similar to that of body structures; the strategy will be highly influenced by performance expectations, the production volume, and intended retail price. Cradles constructed of stamped steel are typical for low-cost, high-volume vehicles while aluminum becomes more common as the performance expectations and intended retail price increase.

Cradle material strategy can also be influenced by the loadpath topology and crashworthiness strategy of the structure. These strategies can sometimes demand specific strength and stiffness attributes of the cradle which can influence its material selection.

1.8 Designing for Structural Efficiency

First of all, it's important to recognize that efficiency is most important at the vehicle level. Although efficiency is often discussed and engineered at the sub-system level, vehicle efficiency is most important, since the automobile is the product the manufacturer is selling. Many different efficiencies are important to a manufacturer, including cost, mass, and BEV driving range efficiency.

$$vehicle\ mass\ efficiency = \frac{vehicle\ mass}{vehicle\ size} \tag{1.1}$$

$$vehicle\ cost\ efficiency = \frac{vehicle\ cost}{vehicle\ size} \tag{1.2}$$

$$vehicle\ range\ efficiency = \frac{driving\ range}{battery\ size(\mathrm{kWh})} \tag{1.3}$$

While some efficiencies are mutually beneficial, some are not. High mass efficiency increases range efficiency, for example, while mass efficiency gained through the application of lighter, more expensive materials is does not necessarily promote cost efficiency.

Designs for all sub-systems and components of an automobile balance to create vehicle-level efficiencies, so it's not surprising that achieving vehicle level efficiency is a complex and difficult process of integrating and balancing vehicle needs.

Vehicle Structure Efficiency

The basis for vehicle structure efficiency is also vehicle size, however there is a stipulation that the design satisfies all *the intended performance. Vehicle structure mass and cost efficiency, for example*:

$$vehicle\ structure\ mass\ efficiency = \frac{vehicle\ structure\ mass\,^*}{vehicle\ size} \qquad (1.4)$$

$$vehicle\ structure\ cost\ efficiency = \frac{vehicle\ structure\ cost\,^*}{vehicle\ size}$$

* design of vehicle structure satisfies all performance loadcases

The process of achieving structural efficiency is systematic, addressing first the influencers that have the largest impact and of which other influencers are dependent on. It also involves consideration of how structure evolves within the vehicle development process, where definition of global topology occurs before local, more granular aspects of the structure solidify. This is often referred to a "coarse to fine" process.

The order in which structural efficiency should be engineered is as follows:

1. *Geometric Efficiency of Loadpaths:* The most significant influencer on structural efficiency is always geometry, so the hierarchy of the structural efficiency process begins with the geometry of the global structural loadpath topology. A firm understanding of how the structure is used in loadcases is required and required early during the vehicle design development process, as spatial constraints begin to form early. Considering that such understanding is required early in the vehicle development process and that this process involves balancing the spatial needs of many sub-systems as well as vehicle level performance, achieving a highly efficient structural loadpath topology is a challenging task and not always achieved.

2. *Minimize Loads:* The magnitude of load being applied to the structure is an obvious influencer on efficiency; the lower the force, the less structure that is required. Note that it is only fruitful to minimize those loads that are driving the structural execution. For example, imagine vertical loads applied to the *front-shock-cap* in two loadcases; 'driving over a speed bump' and 'driving through a pothole'. If the force applied to the structure is higher in the pothole loadcase than in the speed bump loadcase, a suspension change that lowers *shock-cap* loads only in the speed bump loadcase would not allow the body to reduce its strength. Thus, increasing body structure efficiency through lowering loads requires an understanding of which loadcases are 'driving the design' in each area of the structure. Furthermore, minimizing loads often requires a compromise with other vehicle expectations (the suspension tuning of the above *shock-cap* scenario might reduce ride performance, for example), and thus minimizing loads is typically a complex activity requiring a balance of vehicle expectations.

3. *Distribute Loads:* Distributing forces over a larger area fundamentally reduces the maximum force magnitude. This principle can often be leveraged within a structural loadpath to reduce the internal forces of individual loadpath elements.

4. *Section Size and Shape:* The shape and size of each member within the structure has influence on efficiency and can be considered one aspect of 'local geometric efficiency'. Just as with geometric efficiency of the global structural loadpath, the section

size of structural members typically has consequences on other vehicle sub-systems or performance expectations and understanding how influential the shape and size of each member is to efficiency is needed very early in the vehicle design development process.

5. *Local Geometric Features:* Efficient local geometry continues with local geometric features such as strategic local surface positioning and the application of stiffening formations, crush initiators, and lightening holes. The foundation for efficient local design is avoiding abrupt changes in stiffness and strength and strategic placement of surfaces and the application of local stiffening formations. Strength discontinuities are sometimes desirable and geometry is strategically engineered to initiate crush in a crashworthiness loadcase. An example of such geometry can be seen in Fig. 1.33, which shows a feature to initiate *B-pillar* bending in a Side Impact event; further explored in Sect. 4.4.

Lightening holes, those holes which have been incorporated with the primary intention of reducing mass, are very often leveraged to improve the structure's mass efficiency. Identifying locations for lightening holes requires an understanding of local stress and strain levels in several different loadcases and thus computer simulation is often used in the process. Figure 1.33 illustrates these and several other local geometry elements.

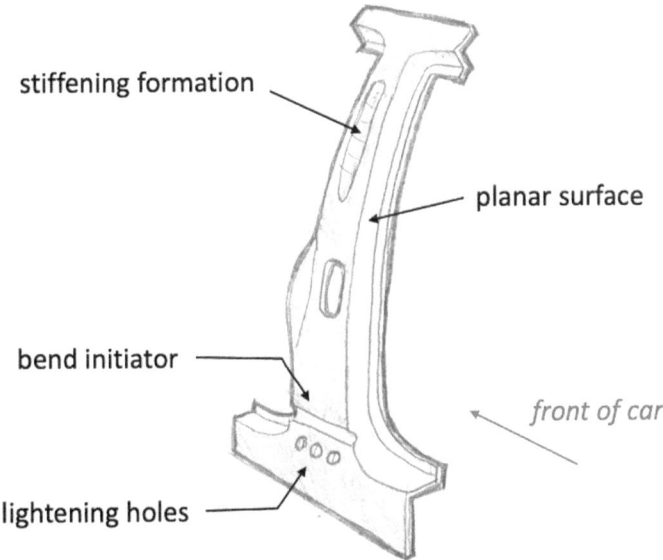

Fig. 1.33 Local geometry examples in a B-pillar-outer-reinforcement

6. ***Material Type Selection:*** A decision to use high strength to weight ratio materials such as aluminum or composites, in the vehicle structure can have significant effect on the structure's mass efficiency, however their usage typically degrades the vehicle's cost efficiency. Hence, the prioritization of geometry over material in this structural efficiency hierarchy.

7. ***Material Grade and Thickness Selection:*** Finally, the selection of material properties and thickness for each individual part impacts mass and cost efficiency. The phrase, "the right material for the right location" is often used to describe this selection process but can extend to the selection of material type too. Understanding how each individual part participates within the structure to satisfy all the performance loadcases is a prerequisite. If an individual part's contribution is primarily stiffness, it is considered a "stiffness dominant" part and thickness selection will be most important. On the other hand, if an individual part's primary contribution is strength, it is considered a "strength dominant" part and its participation and the selection of material grade strength will be most important.

Although this is not a comprehensive list, it should provide an understanding of the hierarchy and the dependency between efficiency enablers; remember, order is important.

It is interesting to note that the efficient solution might look different for different manufacturers. A manufacturer who has an ample supply of welding equipment and Body Shop floor space might find a multi-part construction strategy most cost efficient for the structure while a start-up manufacturer might find it best to assemble fewer, large parts.

Alternatively, a low-volume manufacturer of exotic sports cars might find it most cost efficient to select a composite intensive material strategy while a high-volume manufacturer of entry-level cars might find it most cost effective to choose a steel intensive strategy, as discussed in 1.1.7.

That manufacturer of exotic sports cars might also select a tall rocker section which improves mass efficiency but makes entry into the automobile difficult. That same rocker section would not be acceptable to a manufacturer of a high-volume automobile based on the different customer expectations that designs for.

The most efficient design solution for each car and each manufacturer is influenced by many factors and will not necessarily be the same for all.

How Electric Vehicles Changed the Story

Abstract

There are electric vehicles on the market that have been designed using the time-tested logic and methods of an ICE automobile. We will find in this chapter, however, that an electric vehicle is not simply an ICE automobile with a different powertrain and fuel storage unit. To achieve a successful, efficient EV, component packaging as well as the approach taken for structural development requires change.

The difference in componentry is significant enough to require that the vehicle design development community approach packaging and structural design development differently.

Figure 2.1 illustrates the propulsion content differences between an ICE automobile, a BEV, and a FCEV; where ICE is the internal combustion engine, f is the liquid fuel tank, PB is the propulsion battery, m is the electric motor, and FC is the fuel cell unit.

This chapter assumes that propulsion battery cells have an energy density of roughly 600 Wh/l; typical at the time of this book's writing. Such a density requires BEVs to have approximately 100 kWh of propulsion battery to achieve a 300 mile (480 km) driving range; a range which the industry has determined to be a threshold of customer expectation.

The volume required for a 100 kWh propulsion battery is relatively large and can create challenges for the design development process. It is appropriate to note that there is typically motivation to achieve the driving range expectations with the smallest possible

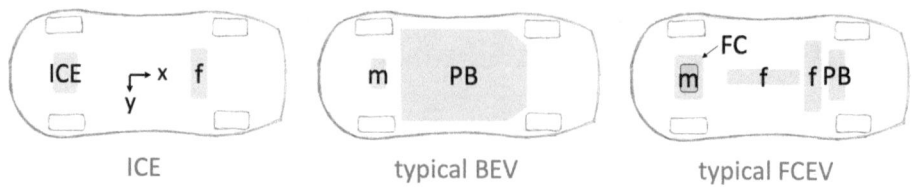

Fig. 2.1 General configuration of propulsion types (front of car to left)

propulsion battery size. Doing so minimizes the cost and the spatial challenges associated with the battery. Aerodynamics and mass have significant influence on range, and it is typical for BEV development to focus on these efficiencies as an enabler for minimizing the battery size.

2.1 Topology

Smaller BEVs whose range is significantly less than 300 miles can package the propulsion battery within traditional automotive structural topologies, such as the 1996 GM EV1, which packaged the propulsion battery in the tunnel and in the lateral space traditionally occupied by the liquid fuel tank (Fig. 2.2a), and the 2008 Tesla Roaster, which packaged its propulsion battery between the occupants and the rear motor (Fig. 2.2b). Both automobiles are small, two-seat configuration with a maximum driving range of 110 and 220 miles respectively.

The larger propulsion battery required for BEVs of a more mainstream configuration and a 300 mile range has been packaged under the center-compartment in nearly all BEVs in this era, from the 2018 Nissan Leaf to the 2022 Ford Lightening. The configuration of these propulsion batteries can be called "underfloor".

Figure 2.3 illustrates the implication to BFI structural topology; particularly the impact to the *center-compartment-mid-rail*. In an ICE, the front of the *center-compartment-mid-rail* is aligned laterally to the *front-compartment-mid-rail* and is a key element of the

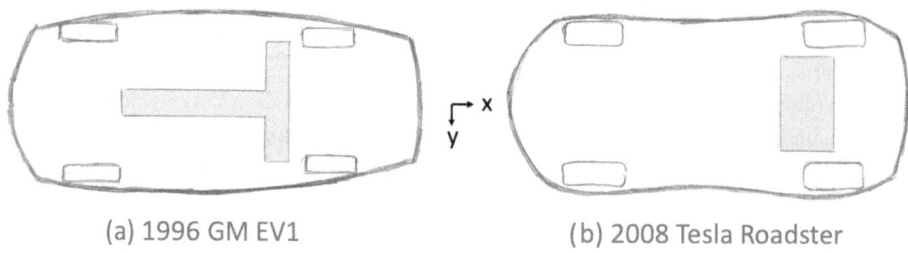

(a) 1996 GM EV1   (b) 2008 Tesla Roadster

Fig. 2.2 Propulsion battery packaging of early BEVs

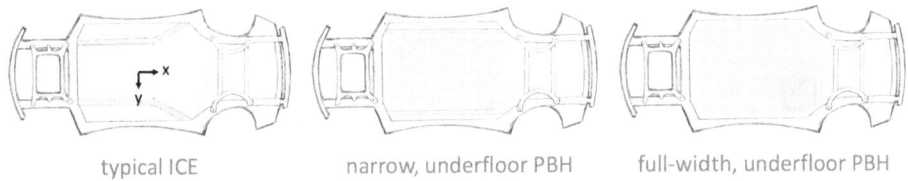

typical ICE   narrow, underfloor PBH   full-width, underfloor PBH

Fig. 2.3 Implications to the center-compartment-mid-rail

loadpath for front crashworthiness and vehicle bending loadcases. Its topology as it moves rearward in the center-compartment varies, stopping just rearward of the *3-bar* in some vehicles while continuing to the *rear-torque-box* in others. BEVs with a narrow, underfloor PB configuration require that the *center-compartment-mid-rail* be moved outboard, creating a misalignment between it and the *front-compartment-mid-rail*, as illustrated in Fig. 2.3b. BEVs with a full-width, underfloor PB configuration require the elimination of the *center-compartment-mid-rails* completely, unless the floor height is exceptionally high.

As *center-compartment-mid-rail* alignment worsens, or the *rail* is eliminated completely, other elements of the structure need to compensate and careful development of the loadpath is required.

Presence of the underfloor propulsion battery can often affect the structural loadpaths positively. It can add stiffness to the center-compartment and boost vehicle-level stiffness, particularly torsional stiffness, as illustrated in Fig. 2.4. The amount of which the PB increases stiffness is dependent on the structural content of the propulsion battery housing, how it attaches to the body structure, and how stiff the body is at those locations. An underfloor propulsion battery can also provide a loadpath in a side crashworthiness loadcase, depending on the structural topology of the battery housing and structural contribution of the battery cells.

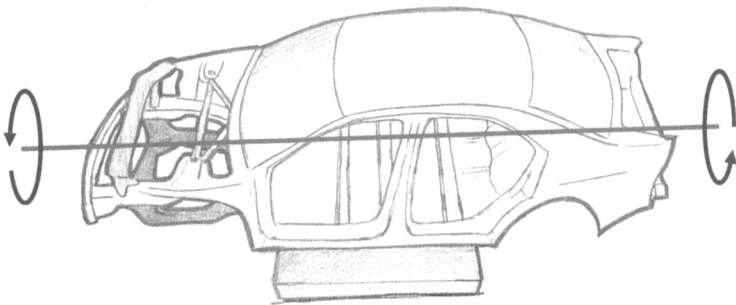

Fig. 2.4 Torsional stiffness contribution of BEVs; PBH and/or PB-cells

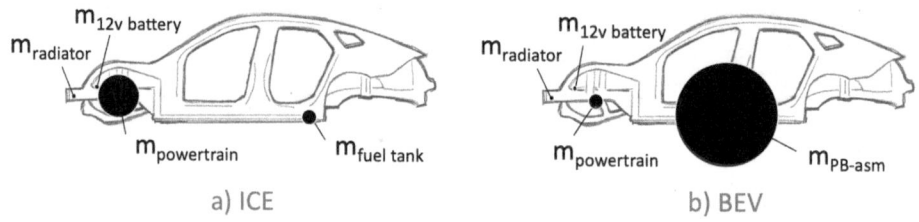

Fig. 2.5 Mass distribution differences

2.2   Mass Distribution

The mass implications of a 100 kWh propulsion battery in the era of 600 Wh/l battery cells is quite significant to vehicle structure and structural performance. As illustrated by Fig. 2.5, the structure is required to manage more mass than an ICE and the mass concentration is quite different. The mass of a BEV with an underfloor PBH configuration is concentrated lower and more towards the center of the vehicle than its ICE counterpart. This concentration affects many loadcases including frontal crashworthiness loadcases and the dynamic bending stiffness of the vehicle structure. This condition also affects vehicle dynamics and suspension tuning, as vehicle's polar moment of inertia is significantly different. Subsequent loadcase specific chapters will cover these effects in more detail.

Figure 2.5 is also a good example of the importance of reviewing a system level FBD. It is often the case that automobile product development focuses on detail behavior of a local area of the structure and consideration of the system level FBD does not occur until the team discovers that their solution concepts are not behaving as they expect. Always remember that behavior of local structure is dependent on the vehicle-level FBD and ensure that you give the FBD appropriate consideration and thought.

2.3   New Structural Considerations

The propulsion system of a typical EV operates at a voltage well above 12 V; 400 V and 800 V are typical at the time of this book's writing. Elements of a EV operating at high voltage include; the propulsion battery, the electric drive motors, and some accessory components. Although mechanisms exist to quickly disconnect the battery from the high voltage circuit, additional safety consideration accompanies the introduction of high voltage content into the automobile. Associated risks include fire resulting from a shorted circuit and electrocution resulting from exposure to live high voltage circuitry.

The need to protect these components motivates manufacturers to increase the size of the structure's 'safety cage' (the portion of the structure that does not incur meaningful plastic deformation during crash events) and package high voltage content within that

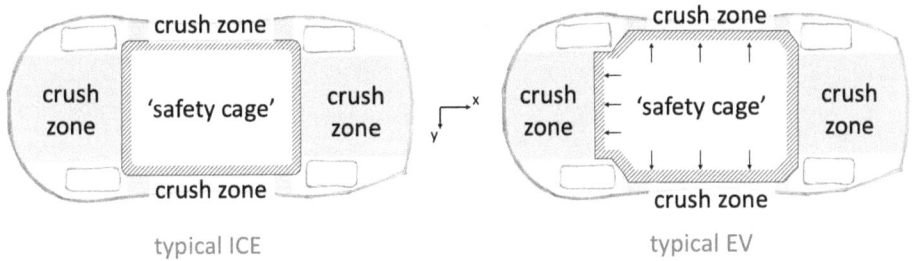

Fig. 2.6 Safety cage zones

zone. Figure 2.6 illustrates an increased safety cage zone. The extended forward portion of the safety cage allows for packaging of high voltage components and the extended side portion allows for protection of the propulsion battery content.

Such adjustments to the size of the safety cage and crush zones can have significant implications to the structural strategy and crashworthiness strategy and consideration in the early stages of the vehicle design development process is critical.

The safety zone/ crush zone relationship is the same for FCEVs, where the hydrogen fuel tank(s) and high voltage components must be protected.

Loadcase Fundamentals

3

Abstract

This chapter provides a scientific foundation for subsequent loadcase chapters as well as a foundation of loadcase terminology. Collision physics will be introduced and the chapter material will build to the construction of a general collision model. It is important to note that the goal of this chapter is not only the derivation of equations. Of equal importance is the reader's development of intuition with regards to collisions and the physics that governs automotive structure, and the information is presented in a way to encourage such development.

3.1 Nomenclature and Terms

Nomenclature

Regulatory Loadcase—A loadcase whose performance must be met in order to sell the vehicle in a particular country or countries.

Consumer Metric Loadcase—A loadcase for which a rating is published for the purpose of providing potential customers with data; to aid their automobile buying decision process.

Due-Care Loadcase—A loadcase that does not otherwise exist but recognized as essential to ensure adequate product performance.

Near-side—the side of the automobile involved in the collusion.

Far-side—the side of the automobile opposite of the collusion.

Ride-down—the act of an object's energy reduction during a collision event.

Curb weight—the weight of the automobile, without occupants and with full fluids (fuel, washer fluid, oil, coolant, etc.).

Curb mass—the mass of the automobile, without occupants and with full fluids (fuel, washer fluid, oil, coolant, etc.).

Corner weight: the downward force being applied by one tire of the automobile.

Load capability: Strength limit or strength behavior as a function of intrusion.

Terms

	Units	Base units
Momentum (p)	N s	kg m/s
Energy (E)	J (N m)	kg m^2/s^2
Work (W)	J (N m)	kg m^2/s^2 (*a form of energy*)
Spring rate (K)		N/m

3.2   Collision Physics Deep Dive

This chapter focuses on building a generic collision model and it gets a little nerdy; understand there is a purpose to it. The physics and concepts developed in this section are the basis for the crashworthiness loadcase sections that follow. Understanding the foundational physics covered in this section will enhance your comprehension of each loadcase. Furthermore, developing a deep understanding of foundational collision physics will reinforce an automotive structural engineer's intuition which, in turn, will enable them to identify successful, innovative, and efficient designs.

One might suggest that such understanding and intuition is not necessary because computer simulation is used during the development process. In truth, computer simulation only produces results, not solutions. An understanding of collision physics allows the engineer to leverage the results; providing insight on how simulation variables and design attributes interact. Without an understanding of foundational collision physics, efficient design solutions are extremely hard to identify.

3.2.1   Applicable Conservation Laws

Energy

Conservation of energy is an important attribute of crashworthiness events and is one to always consider when rationalizing vehicle behavior. It's very useful to compare the total system energy at the end of the collision event to the energy just before. Is there any kinetic energy at the end of the test or has all the event energy been converted into material fracture and crush?

The conservation of energy is simply ...

$$E_o = E_f \qquad [Joules] \qquad (3.1)$$

where:

E_o = the total system energy just prior to the event

E_f = the total system energy after the event

Momentum

Conservation of momentum is another important attribute to consider when reviewing crashworthiness events. Momentum (p) is always conserved in a collision, we find that...

$$p_o = p_f \qquad [Ns] \qquad (3.2)$$

... an expanded form considering that the two collision objects, the 'striking object' and the 'target object', might have an initial velocity...

$$m_T v_{To} + m_S v_{So} = m_T v_{Tf} + m_S v_{Sf} \qquad (3.3)$$

where:

m_T =   mass of the target object

m_S =   mass of the striking object

v_{To} =   velocity of the target object just prior to the collision

v_{So} =   velocity of the striking object just prior to the collision

v_{Tf} =   velocity of the target object after the collision

v_{Sf} =   velocity of the striking object after the collision

Collision models can be an effective tool to visualize and rationalize the high-level vehicle behavior. The following three high-level collision models reflecting the forementioned conservation laws can serve as valuable reference and committing the described physics to memory will help build your structural intuition. Section 3.2.3 will expand upon these models to include the participation of collision object attributes and properties. Although the equations shown in this book can be used to estimate vehicle behavior (especially useful early in a vehicle development process when CAE simulation models are not yet available), their primary intent is to build the reader's understanding and intuition.

The high-level collision model illustrated in Fig. 3.1 represents the condition where both objects are free to move in the direction of impact.

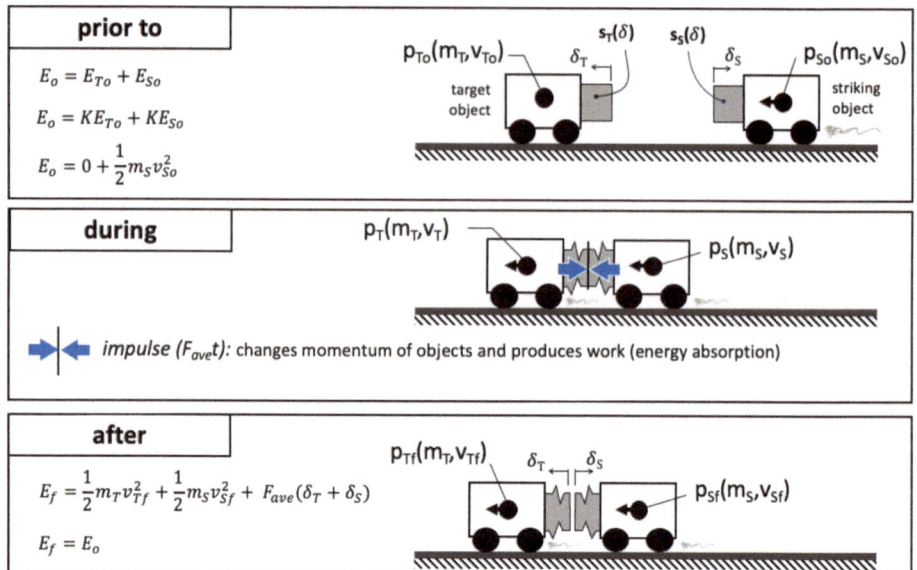

Fig. 3.1 High-level collision model—unconstrained objects

The collision model illustrated in Fig. 3.2 represents the condition where the motion of the target object is constrained. In some cases, the target object has plasticity and crush will occur in it.

The collision model illustrated in Fig. 3.3 represents the condition where translation of the target object is not constrained but is restricted by an external force. The sign of the work performed by the friction might be counterintuitive. The friction acting on the target object is adding resistance to the object's translation and thus adding energy to the collision event. -*note that the equation for E_f assumes that all objects have come to a rest.*

Imagine the difference between pushing two different shopping carts, one with free rolling wheels and one with wheels that are damaged and resist rotation. The resistance provided by the damaged wheels makes that cart harder to push; requiring more effort to move it.

Things to Consider

It is important to recognize any external influences that could have a meaningful contribution to the system energy. For example, a collision where a vehicle collides with the side of a stationary vehicle will involve sliding friction between the struck vehicle's tires and the ground. This frictional force will influence the system FBD and the collision system energy.

Energy is also dissipated in a collision through heat and sound, but these energy paths can be ignored due to their small relative magnitude.

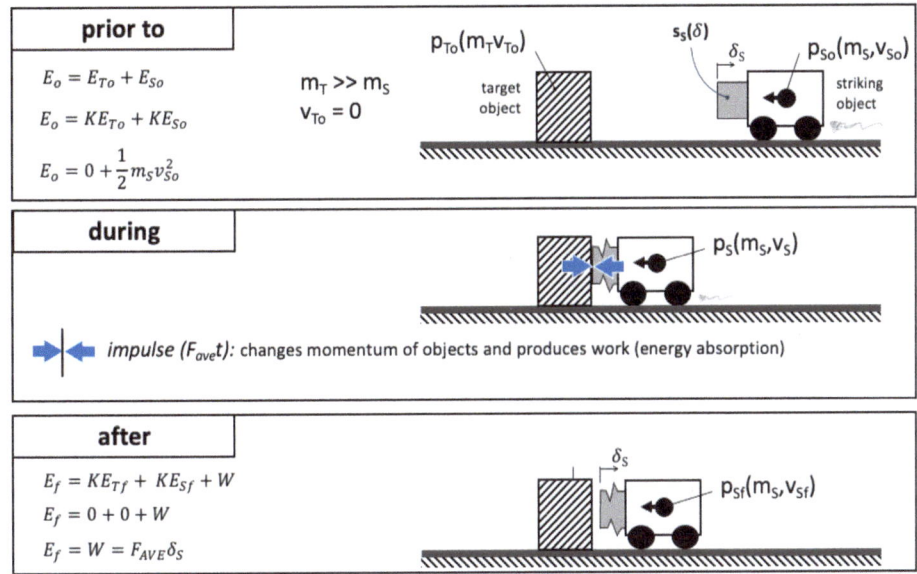

Fig. 3.2 High-level collision model—constrained target object

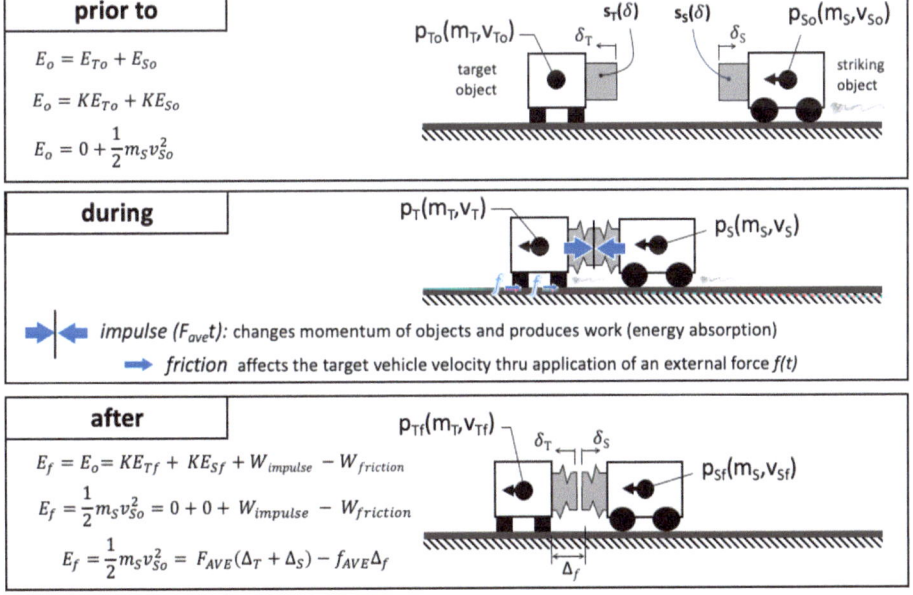

Fig. 3.3 High-level collision model—potentially meaningful external influences

3.2.2 Visualization of a Collision

Per Newton's second law, (F = ma), there is a relationship between an applied force, an object's mass, and its acceleration. It follows that a force must be applied to the object's mass(s) to change its velocity. There are many ways to visualize or experience this notion, including striking an approaching football/soccer ball, as illustrated in Fig. 3.4. Your foot applies a force onto the ball as it collides with it. The force will accelerate the ball and change its velocity (speed and direction).

You might have thought of the scenario where you miskick the ball and strike it on its very side edge. Here, the applied force will induce a rotation but make little change in the ball's direction or speed. This occurs because of a misalignment between the applied force and the ball's center of mass; force was not effectively applied to the object's mass and therefore little change occurred in its velocity.

The fact that a force directed at an object's mass is required to change its velocity is an idea that can be quite elementary and intuitive. So intuitive, unfortunately, that we sometimes fail to recognize it when observing crashworthiness loadcases. Remember this fundamental, as it is key to understanding collisions, building collision models, and is valuable to recognize in engineering practice.

Fig. 3.4 Experiencing Newton's second law

3.2.3   Modeling and Understanding Collisions

Collisions are an isolated event in which a relatively high force acts on the involved objects, over a relatively short amount of time. There are no external forces that act on the objects in a true collision. Collisions can be elastic, in which there is no plastic deformation within the objects and kinetic energy is preserved, or they can be plastic, in which permanent deformation occurs and kinetic energy is <u>not</u> preserved. Automotive crashworthiness events and loadcases certainly fall into the category of plastic.

There are several crashworthiness loadcases wherein the vehicle under evaluation is accelerated due to a collision. An example is the IIHS Side Impact loadcase, in which the test vehicle is stationary and struck by a moving barrier (Fig. 3.5, Link 3.1). In others, the moving test vehicle strikes a stationary barrier and the crashworthiness strategy involves changing the direction of the vehicle's motion; as in Small Overlap Frontal (Fig. 3.6, Link 3.2). The general physics of these tests are the same, however the effect of boundary condition differences apply minor tweaks.

	IIHS-Side Impact test of a 2022 Chevrolet Malibu (https://www.youtube.com/watch?v=KotJsAg43Jw) IIHS YouTube channel, protocol 2	(Link-3.1)
	IIHS-SOF test of a 2021 Tesla Model-Y (https://www.youtube.com/watch?v=5G1dF392iys) IIHS YouTube channel	(Link-3.2)

At a high level, four notable actions are occurring during automotive crashworthiness collisions;

- momentum and velocity of the colliding objects change,
- energy is absorbed elastically and later released,

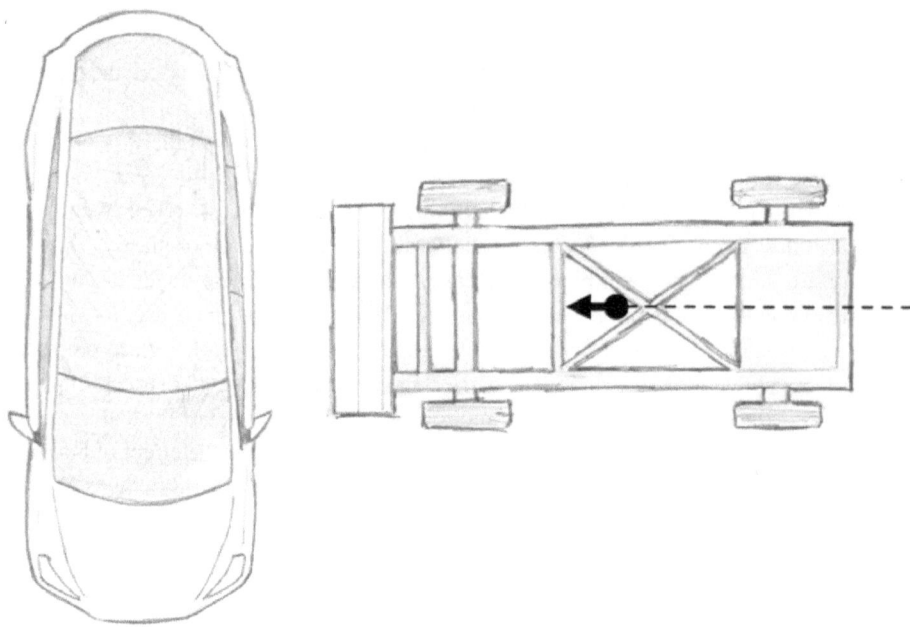

Fig. 3.5 Stationary test vehicle struck by a moving barrier

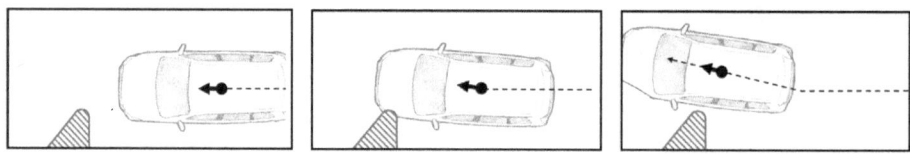

Fig. 3.6 Moving test vehicle striking a stationary barrier

- work is done in the form of plastic deformation, and
- friction acts as an external force, in some cases.

We will build a concise math and mental model by exploring each action independently and then consider a system in which all are present.

3.2.3.1 Rigid Body Elastic Collisions (No Energy Loss)

The model of a collision starts with Newton's second law and the conservation of momentum. Let's consider a collision between a stationary object (the target object) and one with an initial velocity (the striking object), as illustrated in Fig. 3.7.

The momentum of these objects will change during this event; the target object's momentum will increase while the striking object's momentum will decrease. Newton's second law can be written in the form:

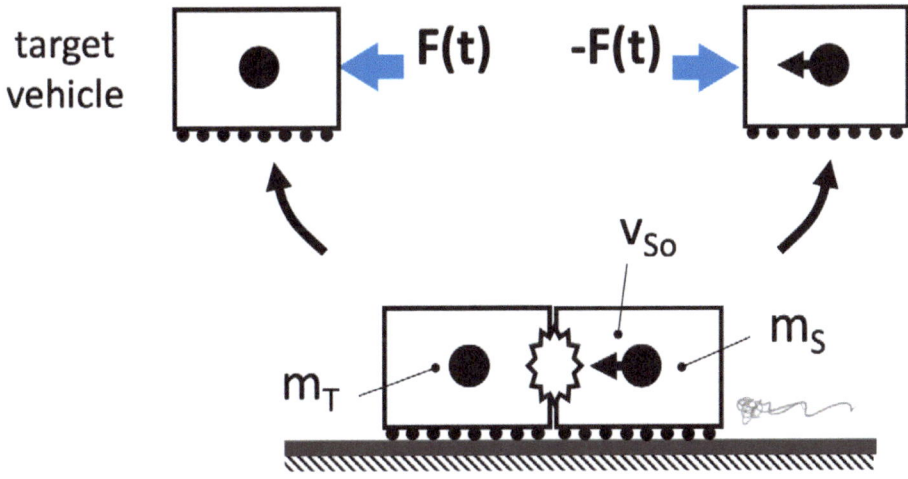

Fig. 3.7   Rigid body elastic collision model

$$dp = F(t)dt \tag{3.4}$$

The change in an object's momentum is impulse (J)…

$$J = p_o - p_f \qquad \text{[N s] or [kg m/s]} \tag{3.5}$$

$$\text{Impulse: } \boxed{J = \int_o^f F(t)dt} \tag{3.6}$$

or

$$\boxed{J = F_{ave}\Delta t} \tag{3.7}$$

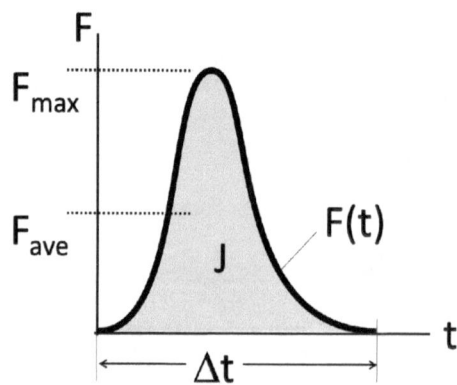

where:

J = impulse

p = momentum of a collision object

F = force

t = time

The relationship shown in Eq. 3.7 can also be derived differently, starting with the equation for acceleration:

$$a = \frac{\Delta v}{\Delta t} \tag{3.8}$$

$$a = \frac{(v_{Tf} - v_{To})}{(t_f - 0)} \tag{3.9}$$

$$a = \frac{\Delta v_T}{\Delta t} \tag{3.10}$$

$$\Delta v_T = a(\Delta t) \tag{3.11}$$

...introducing a force term can be done by using Newton's second law...

$$F = ma \tag{3.12}$$

$$a = \frac{F}{m} \tag{3.13}$$

...substituting into Eq. 3.11 yields:

$$\boxed{m_T \Delta v_T = F_{ave} \Delta t} \tag{3.14}$$

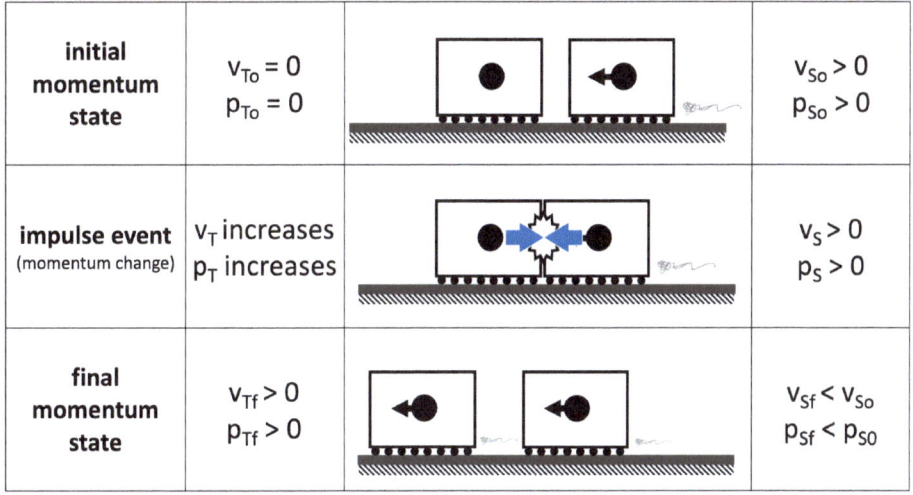

initial momentum state	$v_{To} = 0$ $p_{To} = 0$			$v_{So} > 0$ $p_{So} > 0$
impulse event (momentum change)	v_T increases p_T increases			$v_S > 0$ $p_S > 0$
final momentum state	$v_{Tf} > 0$ $p_{Tf} > 0$			$v_{Sf} < v_{So}$ $p_{Sf} < p_{So}$

Fig. 3.8 Momentum states during a collision

where:

a	=	acceleration of the target object
m_T	=	mass of the target object
Δv	=	target object's velocity change
F_{ave}	=	average collision force
Δt	=	collision time duration

With the understanding that impulse is equal to $m \cdot \Delta v_f$ (Eq. 3.5), we see that Eqs. 3.7 and 3.14 are the same.

Consider the time steps of a collision shown in Fig. 3.8 and notice the change in each object's momentum. The target object's momentum increases and the striking object's momentum decreases as the event progresses.

Impulse might feel like Work, but it is not. The units for Work are $(kg\ m^2/s^2)$ while the units for impulse are $(kg\ m/s)$. **Impulse, or the application of force over a time duration, is the mechanism for an object's momentum change** (Eqs. 3.7 and 3.5).

3.2.3.2   Flexible Body, Perfectly Elastic Collisions

The next step in building our mental/math model is to consider that the colliding objects in automotive crashes are not rigid. To explore the effect of object elasticity, consider a man so fascinated with marshmallows that he wears a belt of marshmallows to an ice skating rink, as illustrated in Fig. 3.9.

Fig. 3.9 Colliding ice skater analogy

There, his friend approaches with speed and pushes the marshmallow man's belt. The end result, of course, is that the man has been accelerated and his velocity has increased. Let's now focus on the marshmallow, which was in the loadpath of this collision event and acted as a spring. Let's explore the forces at the time steps illustrated in Fig. 3.10.

A force is generated between the two skaters while they're in contact and compresses the elastic marshmallow. The force magnitude is influenced by the spring rate of the marshmallow, as the marshmallow can only transmit the level of force corresponding to its elastic modulus. It is only when the marshmallow is fully compressed that it can transmit the entire force that the friend is trying to apply.

If the fact that the spring rate limits the impulse force is not intuitive, try the simple experiment illustrated in Fig. 3.11; place a sponge, piece of foam, or other springy object between your hand and a wall. Try move the wall by pushing on the springy object. As you try, be mindful of how hard you push during the experiment. You will notice that you are not able to exert your full effort until your springy object is compressed. You personally experienced point-A noted in Fig. 3.11b.

Since we are assuming that the marshmallow behaves as an ideal spring and does not undergo plastic deformation, the energy temporarily stored in the marshmallow is released as the impulse force is removed. The release of this energy is seen as an equal and opposite force applied to the collision objects (Fig. 3.10c), working to increase the final velocity of the target object and reduce the velocity of the striking object.

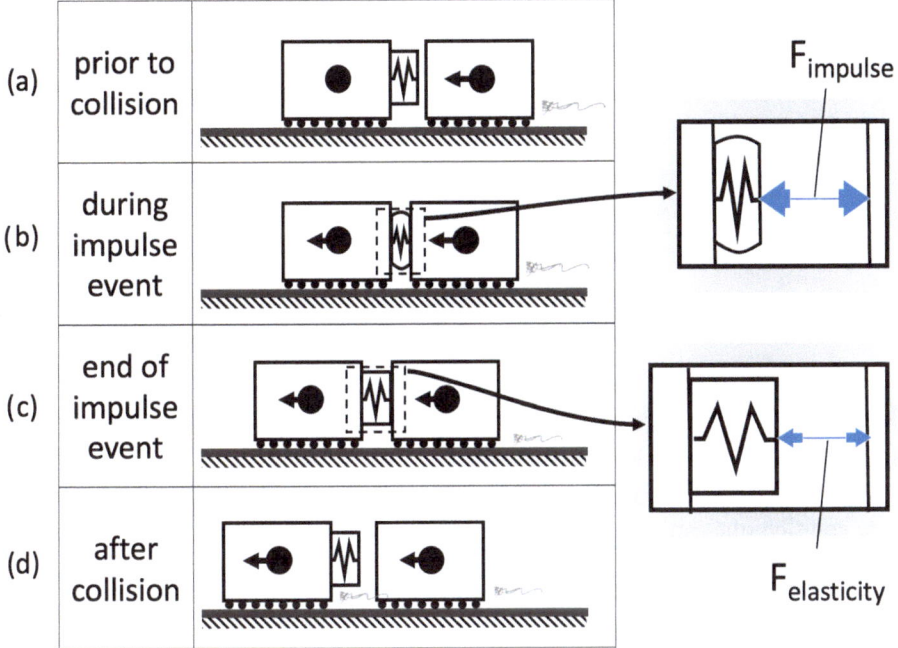

Fig. 3.10 Ice Skater collision time steps

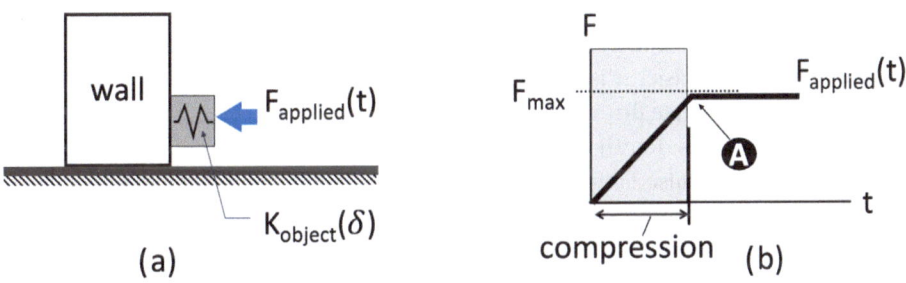

Fig. 3.11 Spring—wall experiment illustrations

This effect is difficult to visualize in the skating marshmallow man analogy as the spring rate of the marshmallow is extremely soft. The man himself has a spring rate, although this is also too soft to visualize release of the stored energy.

Therefore, consider the collision between a pogo-stick rider and the ground. This is an ideal example, as a pogo-stick is a device that is designed to store and release as much of the collision energy as possible. In this analogy, the pogo-stick and rider can be considered the striking object and can be represented by the model illustrated in Fig. 3.12.

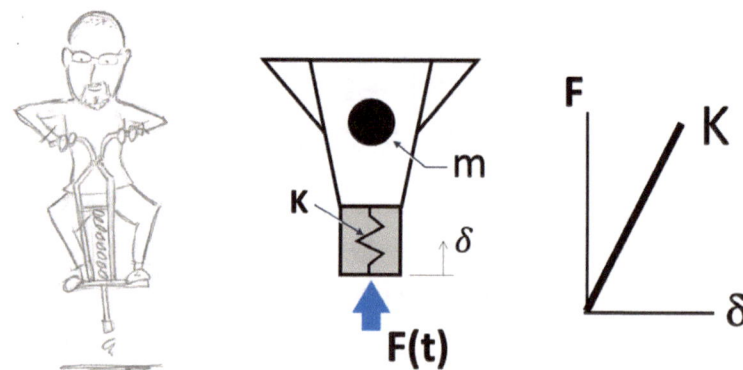

Fig. 3.12 Pogo-stick and rider model

Note that the mass of the pogo-stick and rider is much, much less than the other object in the collision (the earth) and the velocity change of the earth is therefore unfathomably small (velocity change based on object mass is covered later in this in section). Table 3.1 illustrates the timesteps of this collision. Notice how the impulse force is limited by pogo-stick's spring rate and that the pogo-stick absorbs and later releases the energy stored by the spring compression.

If you've ridden a pogo-stick, you'll know that the release of the energy in the spring is not enough to return you to the height at which you started from; there are energy losses in this system. To ride a pogo-stick, the rider must manipulate their body motion to compensate for this energy loss; the better the pogo-stick is at storing and releasing collision energy, the easier it is to ride.

It's appropriate to note that the magnitude of F_{max} is not an independent of the collision attributes and variables. Recall that Eq. 3.6 tells us F_{max} is a function of the impulse event duration and that impulse is the area under the F·t curve. Table 3.2 illustrates how F_{max} changes based on the object's spring rate. Note that we have assumed that the spring rate of the ground to be much stiffer than the pogo-stick's spring; so stiff that we can ignore its effect in the collision. We have also assumed that the pogo-stick's spring is long enough that it never compresses completely. There is certainly the possibility that the stiffness of both collision objects will need to be considered in a math or mental model. The relevance will need to be considered separately for each collision scenario.

Final Velocity of Collision Objects

You'll likely recall from your undergraduate physics course that the final velocity of the objects involved in a rigid-body collision depend on their relative masses. This is, of course, dictated by the conservation of momentum; the total momentum just before the collision is equal to the momentum just afterwards.

Table 3.1 Pogo-stick collision time steps

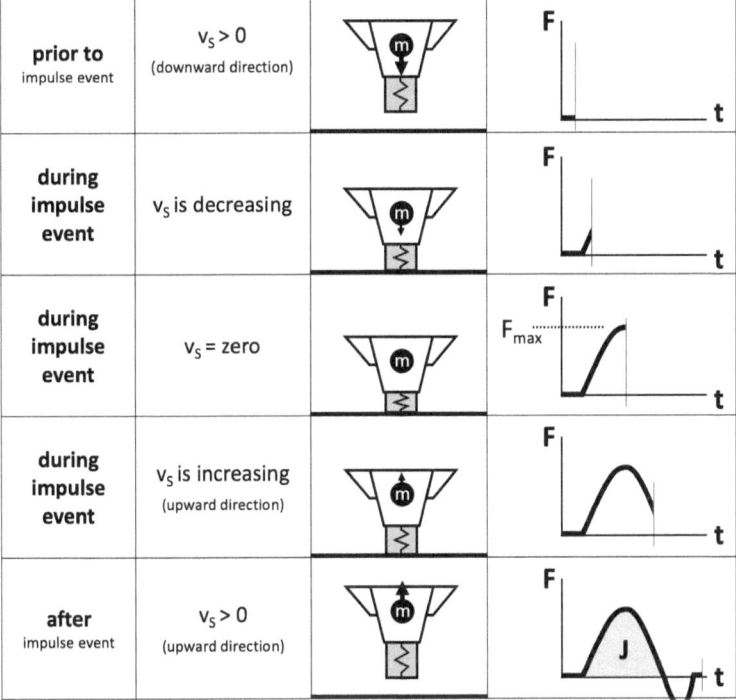

Table 3.2 Relationship between spring rate and F_{max}

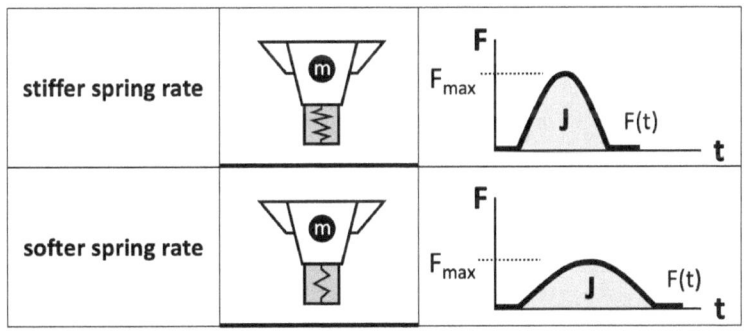

$$p_o = p_f \qquad [\text{kg m/s}] \qquad\qquad (3.15)$$

$$m_T v_{To} + m_S v_{So} = m_T v_{Tf} + m_S v_{Sf} \qquad\qquad (3.16)$$

looking at this equation from the perspective of the target object's final velocity we get...

$$v_{Tf} = \frac{m_T v_{To} + m_S v_{So} - m_S v_{Sf}}{m_T}$$ (3.17)

...and when the initial velocity of the target object is zero, we get...

$$v_{Tf} = \frac{m_S v_{So} - m_S v_{Sf}}{m_T} \quad \text{where } v_{T_o} = 0$$ (3.18)

A useful property of perfectly elastic collisions is that kinetic energy is also conserved.

$$KE_o = KE_f$$ (3.19)

$$\frac{1}{2} m_T v_{To}^2 + \frac{1}{2} m_S v_{So}^2 = \frac{1}{2} m_T v_{Tf}^2 + \frac{1}{2} m_S v_{Sf}^2$$ (3.20)

...with a stationary target object, $v_{To} = 0$...

$$\frac{1}{2} m_S v_{So}^2 = \frac{1}{2} m_T v_{Tf}^2 + \frac{1}{2} m_S v_{Sf}^2$$ (3.21)

We can learn something about the relationship between object final velocities and object masses by using equations developed from the conservation of momentum and the conservation of kinetic energy...

...simplifying Eq. 3.16...

$$m_T v_{Tf} = m_S \left(v_{So} - v_{Sf} \right)$$ (3.22)

...manipulating Eq. 3.21...

$$m_T v_{Tf}^2 = m_S \left(v_{So}^2 - v_{Sf}^2 \right)$$ (3.23)

$$m_T v_{Tf}^2 = m_S \left(v_{So} - v_{Sf} \right)\left(v_{So} + v_{Sf} \right)$$ (3.24)

...dividing Eq. 3.24 by Eq. 3.22 and applying more algebra yields...

$$v_{Sf} = v_{So} \frac{m_S - m_T}{m_S + m_T}$$ (3.25)

...and...

$$v_{Tf} = v_{So} \frac{2m_S}{m_S + m_T}$$ (3.26)

Using Eqs. 3.25 and 3.26 we can project the behavior of three specific cases. Consider the cases shown in Table 3.3, where the white sphere is a ping pong ball and the dark sphere is a golf ball; the golf ball being much, much heavier than the ping pong ball.

Table 3.3 Object behaviors based on object mass

$m_T \ll m_O$	$m_T = m_S$	$m_T \gg m_S$
$v_{Sf} \approx v_{So}$ $v_{Tf} \approx 2v_{So}$	$v_{Sf} = 0$ $v_{Tf} = v_{So}$	$v_{Sf} \approx - v_{So}$ $v_{Tf} \approx v_{So}\,(2m_S/m_T)$
The striking object loses little velocity. The final velocity of the target object vehicle is nearly twice that of the striking object's initial velocity	All KE is transferred to the target object. The striking object is stationary at the end of the event. The target object has the same speed as the striking object's initial velocity.	The striking object bounces off the target object. Its velocity is only slightly reduced but its direction has changed. The target object moves slowly away from the collision.

Although these behaviors assume elastic collisions, they will provide a good starting point for behavior in plastic body collisions.

It is easy to visualize marshmallow elasticity but understand that the man also has elasticity; K_1 and K_2 respectively, as illustrated in Fig. 3.13. This is true of automobile structures as well. Although elastic deflection within automotive structures can be extremely small and difficult to measure, it does exist. Nothing is truly rigid. Elastic deflection within a structure is more significant in high loading events; recall that deflections are proportional to the applied force, per equation $F = K\delta$.

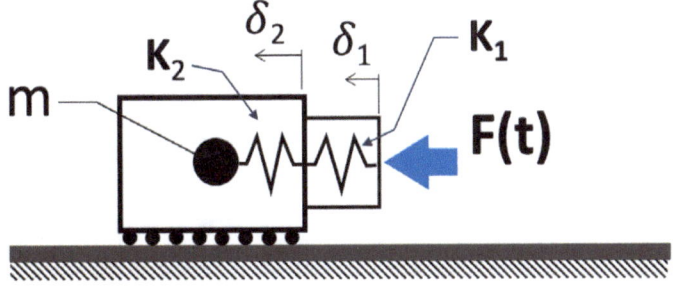

Fig. 3.13 Marshmallow man FBD

3.2.3.3 Plastic Body Collisions

Crush, fracture, and/or general material failure are a big part of automotive collisions and we'll find that including it has significant implications to our math and mental collision models. To represent plasticity, we will consider a target object that has a strength profile, shown as $(s(\delta_p))$ in Fig. 3.14, such that there might be material failure when a force is applied.

Effect on the impulse event

For an analogy, consider the candy-filled piñata shown in Fig. 3.15. Its strength changes as crush occurs in its structure, which can be represented in graphical form, as illustrated in Fig. 3.16; a strength profile as a function of intrusion.

Based on this strength profile, the piñata has the ability to manage force as a function of intrusion, as illustrated in Fig. 3.17. This is called 'load capacity' $(C_{piñata}(\delta))$.

Visualize the collision scenario where a child strikes the piñata with a stick as it hangs from a very long string. The piñata is initially stationary and free to move sideways (the string is long enough that we can ignore the centering effect that the piñata's mass induces as the string changes angle). At first contact, the piñata crushes due to the applied force and little change occurs to its velocity. At some point, however, the piñata stops crushing and its velocity increases substantially. …it would be wrong of me to leave out the fact that candy falls to the ground through the cracks, as that is the point of a piñata. We will ignore this glorious event and assume that the piñata's mass does not change.

The time steps illustrated in Table 3.4 show how F(t) is initially limited by the piñata's force capacity (Table 3.4a). At some point, the force capacity is strong enough to manage the maximum force of the stick (Table 3.4b).

It's appropriate to note that the magnitude of F_{max} is not independent of the collision attributes. Recall that Eq. 3.6 tells us that F is a function of the impulse event time duration and that impulse is the area under the F·t curve. Table 3.5 illustrates how F_{max} changes based on the object's load capacity or crush magnitude. Here we see that, assuming that

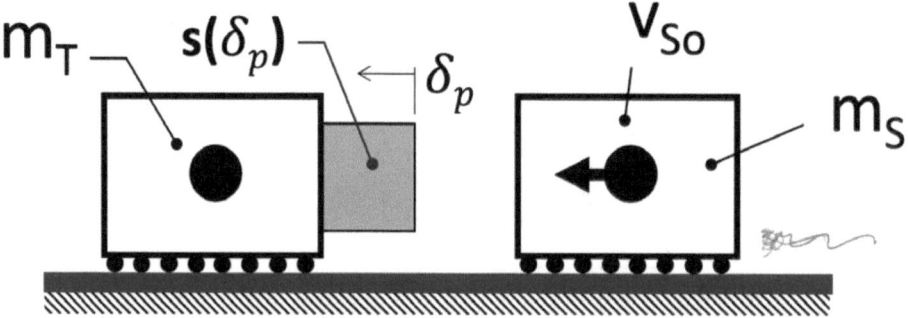

Fig. 3.14 Plastic collision parameters

Fig. 3.15 Piñata analogy

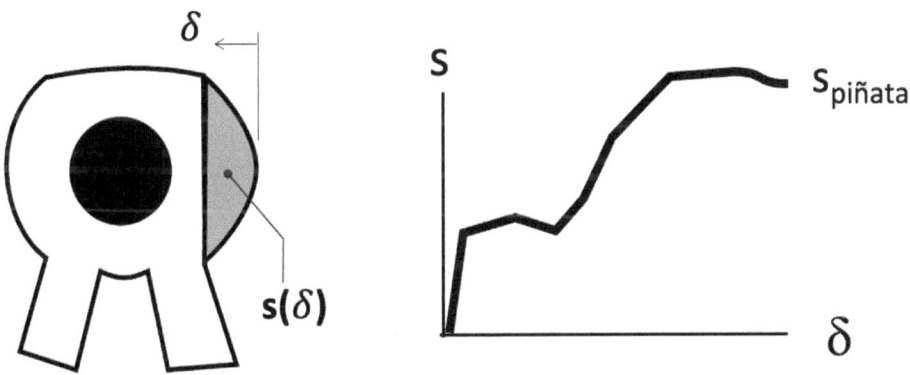

Fig. 3.16 Piñata strength

an equivalent impulse magnitude is required to move the two piñatas, the impulse event time duration increases and the maximum force decreases as the piñata's load capacity decreases.

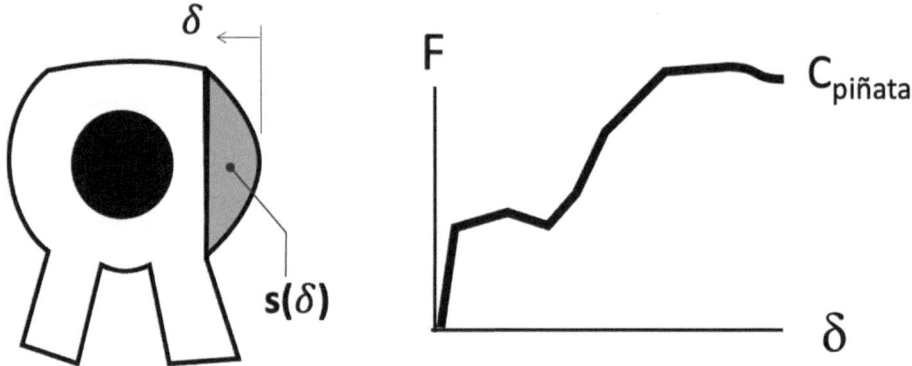

Fig. 3.17 Piñata load capacity

Table 3.4 Stick-Piñata collision time steps

prior to collision		
during impulse event		(a)
during impulse event		(b)

Table 3.5 Relationship between strength and crush/impulse time duration

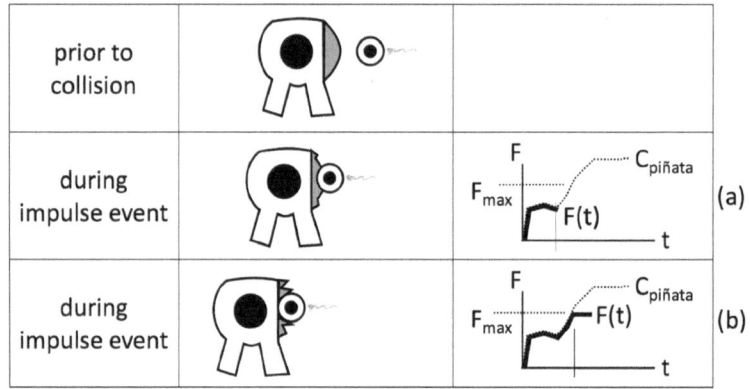

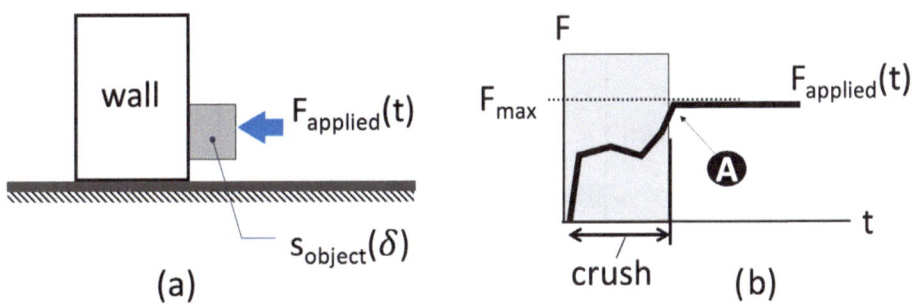

Fig. 3.18 Cup—wall experiment illustrations

The relationship between an object's load capacity and the impulse force is not necessarily intuitive, so an experiment similar that performed for elasticity can be done to reinforce this concept. Consider the following experiment, as illustrated in Fig. 3.18: place a paper cup, cardboard box, or other expendable and easily crushable object between your hand and a wall. Consider the wall and object as a single entity with a force capacity $C_{wall/object}(\delta)$ Now try move the wall/object entity; push and be mindful of how hard you are pushing during the experiment. You will notice that you were not able to exert your full effort until your crushable object was crushed. You experienced point-A noted in Fig. 3.18b.

The time steps in Table 3.6 illustrate a collision involving plastic deformation. It is different from the elastic collision in that compression seen during the impulse event does not spring back to its initial state.

Conservation of Energy

Momentum and kinetic energy have so far been conserved, as we have previously considered only rigid body and perfectly elastic collisions. Momentum is always conserved, and it is true for a plastic body collision. Kinetic energy, however, is not conserved in a plastic body collision. A visual example of that can be seen in Fig. 3.19; the difference between a collision involving a solid rubber piñata (elastic body) and a paper piñata (plastic body). Here we see that some amount of **kinetic energy is absorbed in a plastic body collision.**

The amount of energy absorbed through crush, fracture, and/or general material failure is equivalent to the Work done during the impulse event. …or the vast majority of work done there. Energy is also absorbed through heat and sound, but the energy levels associated with these outlets are insignificant compared to that of material fracture.

We can relate the Work to attributes of the event by starting the model illustrated in Fig. 3.20 and equating Work to ΔKE…

$$W = KE_o - KE_f \text{ [Joules] or [Nm]} \tag{3.27}$$

…Work is also defined as…

Table 3.6 Plastic body collision time steps

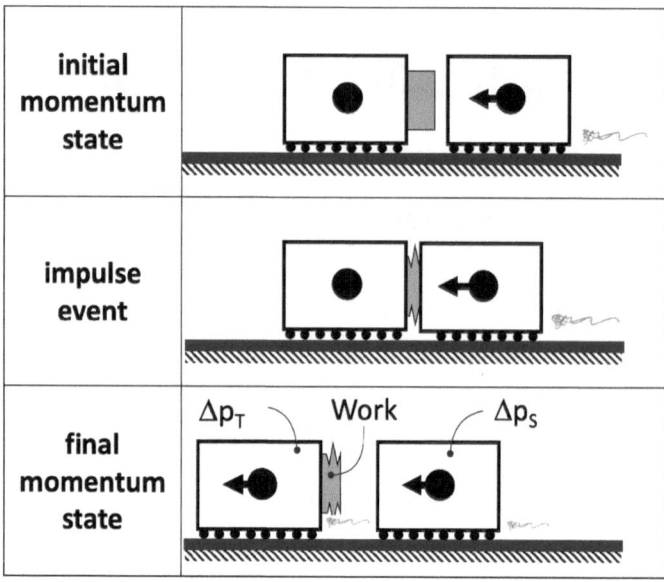

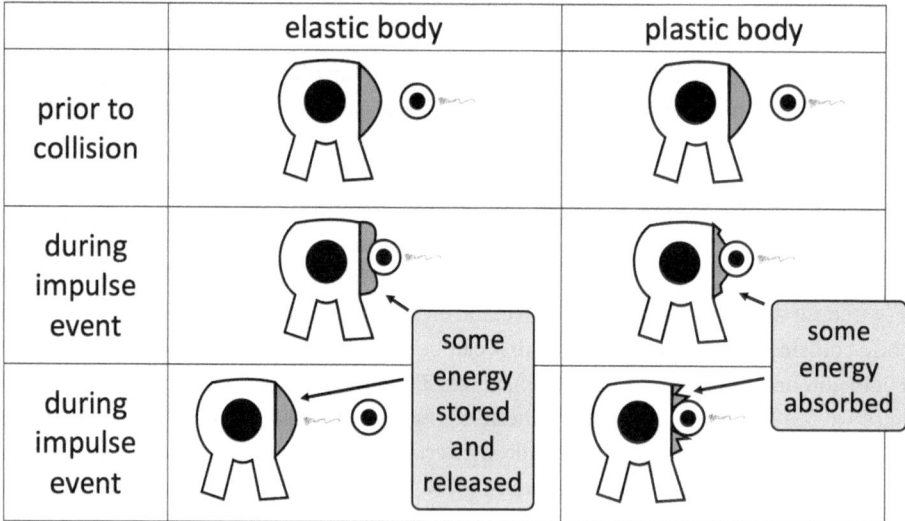

Fig. 3.19 Elastic versus plastic body collision

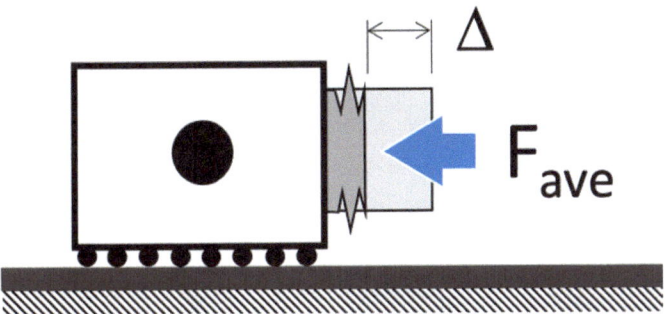

Fig. 3.20 Work model

$$W = F_{ave}\Delta \tag{3.28}$$

...substitution produces...

$$\boxed{F_{ave}\Delta = KE_o - KE_f} \tag{3.29}$$

...further substitution provides:

$$\boxed{F_{ave}\Delta = \left(\frac{1}{2}m_T v_{To}^2 + \frac{1}{2}m_S v_{So}^2\right) - \left(\frac{1}{2}m_T v_{Tf}^2 + \frac{1}{2}m_S v_{Sf}^2\right)} \tag{3.30}$$

Equation 3.30 equates the change in kinetic energy to the average impulse force and the intrusion distance. The understanding that the kinetic energy lost during the event will express itself as crush is an important foundation for our math and mental crashworthiness models.

The piñata examples considered that a strength profile and a crushable element only exists on the target object. Understand that, in many crashworthiness loadcases, this exists on both the target and striking objects. This, however, does not add too much complication.

Also note that we are assuming that the mass of the objects do not change during the event; the mass of any components that fall off of a vehicle during the collision should have an insignificant effect.

3.2.3.4 Friction as an External Force

Automotive crashworthiness events usually don't occur in a frictionless environment; sometimes on ice (as in the marshmallow man example) but crashes are extremely rare while hanging from a string (as in the piñata example). In some cases, friction will be an insignificant force relative to the crash forces and can be ignored. In others, friction will have significant influence on the crash behavior and need to be considered. Recall that the equations built thus far assume that no external forces act on either object. Therefore, if

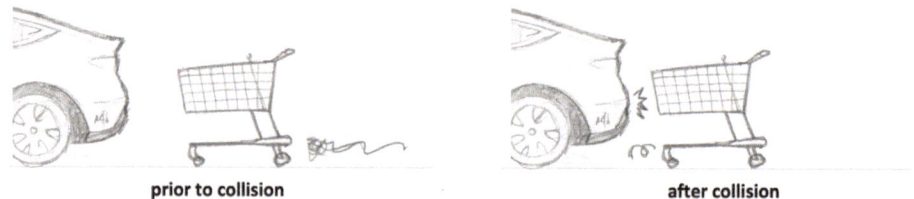

prior to collision   after collision

Fig. 3.21 Stray shopping cart collision

we decide that friction needs to be considered in a loadcase, we must use our engineering sense to adjust our model of the event. This section will explore the relative importance of friction and provide an example of how we might adjust our model.

We will start with a familiar event in which a stray shopping cart crashes into a parked car, as illustrated in Fig. 3.21.

The impact of the shopping cart does not create any forward motion of the car. You might be able to visualize that the collision would not induce velocity in the car even if the car's transmission was in neutral and no parking brake was applied; seemingly violating the conservation of momentum and condition described in Table 3.3. Friction is, of course, the culprit. This is intuitive if you have ever tried to push an automobile in neutral with its parking brake off (Fig. 3.22a). The rolling resistance of the tires and the internal friction of the driveline are quite significant in this case. You can move the automobile but it can take a good deal of force to do so.

Compare that, however, to an attempt to push the same automobile sideways (Fig. 3.22b). The frictional force is significant in this case too. Here we see that the magnitude of the frictional force acting on the tires is much larger than the frictional force(s) occurring in the fore-aft direction. It requires much more force to push the car sideways and we're very unlikely to do so.

Like these two examples, determining if frictional forces are significant to a loadcase requires consideration of the magnitude of the frictional force(s) and the applied force(s). CAE simulation studies can be used to help this determination, although sometimes it requires subjective engineering judgement.

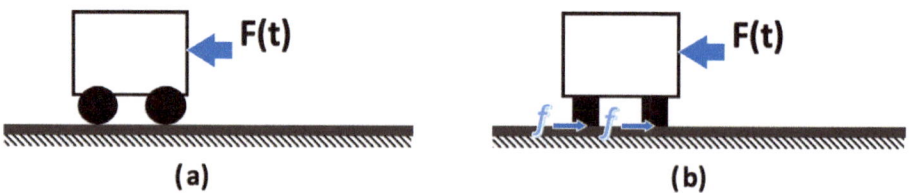

(a)   (b)

Fig. 3.22 Pushing an automobile scenarios

Including Friction

Including friction in our model requires us to be cognizant of two things; the magnitude of the frictional force and the loadpath between the applied force(s) and the frictional force(s).

As you might recall from your introductory physics course, the magnitude of the frictional force (Fig. 3.23) is simply a function of the normal force created by the object and the surface frictional coefficient, as shown in Eq. 3.31.

$$f \leq \mu N \tag{3.31}$$

where:

f	=	frictional force
μ	=	static friction coefficient
F_{CW}	=	corner weight
N	=	normal force

Recall that the friction resists motion with a force up to μN, hence the $\leq$ symbol usage in Eq. 3.31. Once the static frictional force has been overcome, a frictional force associated with a dynamic sliding friction coefficient continues to resist motion.

Recognition of the loadpath between the applied force(s) and the frictional force(s) is important aspect of considering friction, as illustrated in Fig. 3.24. Just as we saw the spring rate of the loadpath between applied force(s) and an object's mass to be important to momentum transfer in Sect. 3.2.3.2, the spring rate of the loadpath between the applied force(s) and the resistive frictional force(s) is important. The stiffness of this loadpath will influence how quickly the object can overcome the static frictional force(s).

If this is not intuitive, try the following experiments with a heavy book laying on a table, as illustrated in Fig. 3.25. Find a marshmallow or other soft, springy object. In the first experiment, place the springy object between your hand and the book and slowly apply force in an attempt to move the book sideways. You will notice that you cannot

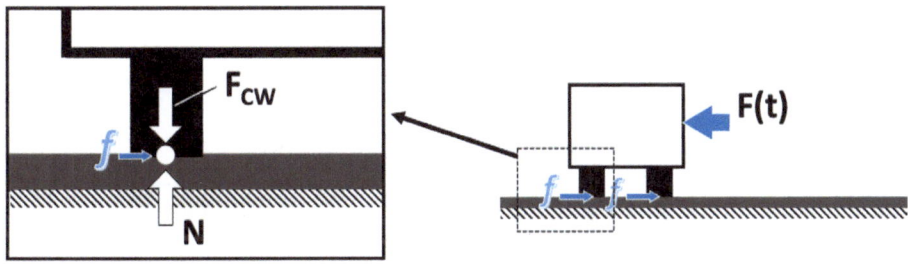

Fig. 3.23 Friction force model

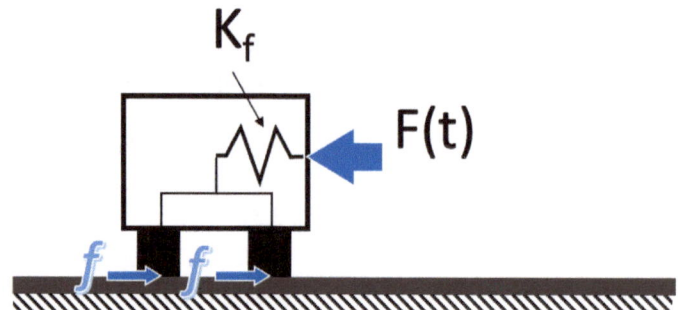

Fig. 3.24 Fictional force loadpath model

overcome the frictional force until there is some compression of your springy object and that compressing the object took some amount of time. For the second experiment, replace the springy object with a rock or something very stiff. You will notice that you overcome the frictional force almost immediately (A. in Fig. 3.25). The amount of time required to overcome the static frictional force is dependent on the weight of the book and the spring rate of your springy object.

We will see that the amount of time that frictional forces resist motion can be significant, as we examine crashworthiness loadcases later in this book. An understanding of the frictional force magnitude and the effect of the loadpath between the applied force(s) and frictional force(s) will become critical.

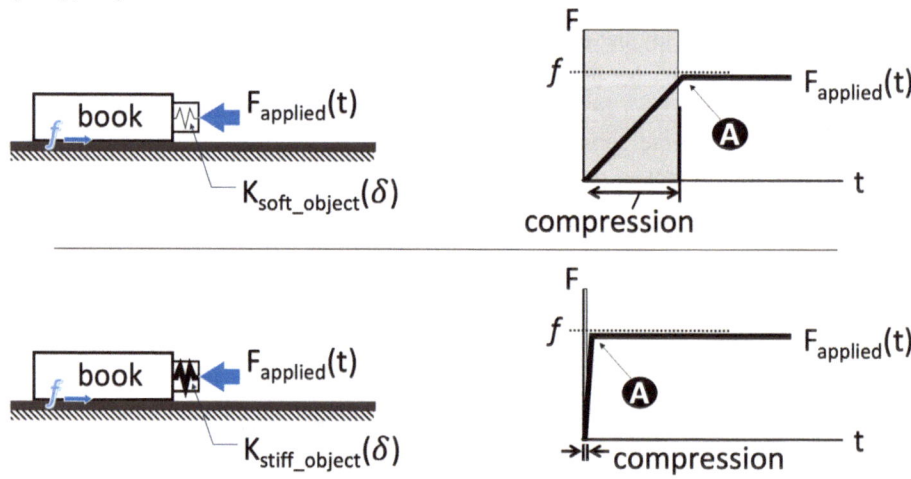

Fig. 3.25 Book—table experiment

3.2.3.5   A Comprehensive Collision Model

We will construct a comprehensive collision model considering all of the elements discussed in the previous sections; momentum transfer, elasticity, plasticity, and friction. Figure 3.26 shows a schematic where all of these things are relevant while the schematic in Fig. 3.27 represents a condition where friction was determined to be insignificant.

Considering a comprehensive collision model starts with the recognition that there is a certain F·t required to accelerate the target object and build its velocity. The component of F·t that is always required is associated with changing the momentum of a colliding object, or the impulse (J), as described in Sect. 3.2.3.1 and defined by Eqs. 3.5, 3.6, and 3.7 (duplicated here as 3.32, 3.33, and 3.34).

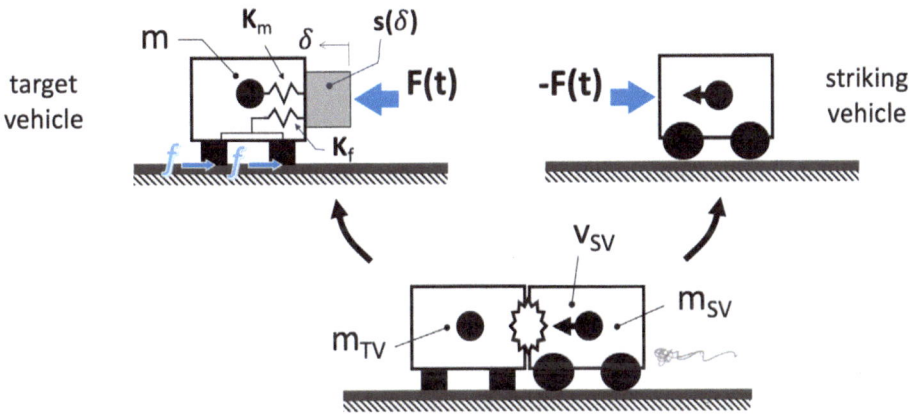

Fig. 3.26 Collision model including elasticity, plasticity, and friction

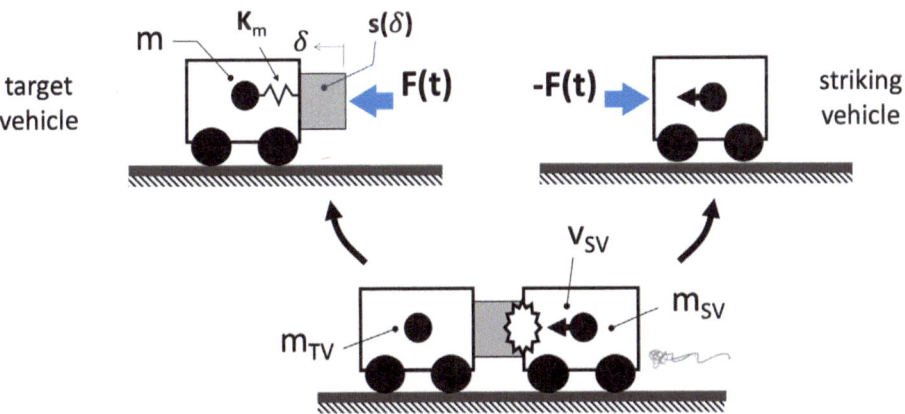

Fig. 3.27 Collision model including elasticity and plasticity

$$J = p_f - p_o \qquad \text{[N s] or [kg m/s]} \tag{3.32}$$

$$\text{Impulse}: J = \int_o^f F(t)dt \tag{3.33}$$

$$J = F_{ave}\Delta t \tag{3.34}$$

where:

J	=	is the impulse
p	=	is the momentum of a collision object
F	=	force
t	=	time

Recall that impulse is not work. It is the mechanism for changing the momentum of a colliding object. The magnitude of impulse is equal to the area under the F(t) curve or the product of F_{ave} and Δt, as illustrated in Fig. 3.28.

The study shown in Table 3.7 explores a relationship between the mass of the colliding objects and their Δp. Perhaps it's intuitive but the study shows that Δp increases as the total mass increases. Since impulse is defined as the Δp and equivalent to $F_{ave}\Delta t$, the conclusion is that a larger impulse magnitude is required as the combined mass of colliding objects increases.

With regards to the shape of the F(t) curve, recall that both **object elasticity** (K_m) and the object's **strength profile** ($s(\delta)$) influence how quickly force builds within the target object; softer and/or weaker target objects delay the force progression, as illustrated in Fig. 3.29. Thus, the shape of the F(t) curve within the t_1 timeframe is dependent on the object's elasticity and strength profile.

Fig. 3.28 Graphical representation of impulse

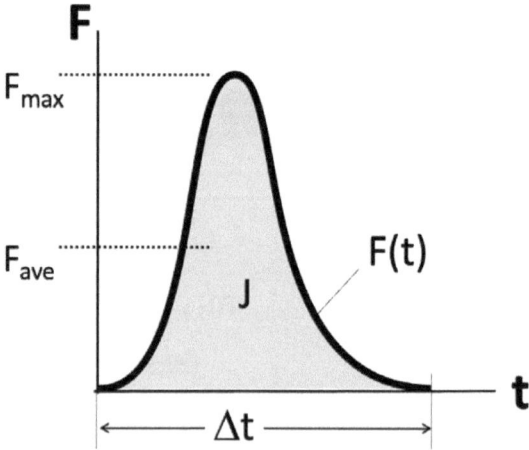

Table 3.7 Effect of mass on impulse magnitude

striking object			target object				conservation of momentum check
mass 1900 Kg			**mass** variable Kg				
V_{So} 16.7 m/s			V_{To} 0 m/s				
V_{Sf}	p_{Sf}	Δp	m_T	V_{Tf}	p_{Tf}	Δp_T	$p_{Tf} + p_{Sf} - p_{So}$
[m/s]	[Kg m / s]	[Kg m / s]	[kg]	[m/s]	[Kg m / s]	[Kg m / s]	[Kg m / s]
2.0	3725	-27941	1500	18.6	27941	27941	0
1.4	2714	-28952	1600	18.1	28952	28952	0
0.9	1759	-29907	1700	17.6	29907	29907	0
0.5	856	-30811	1800	17.1	30811	30811	0
0.0	**0**	**-31667**	**1900**	**16.7**	**31667**	**31667**	0
-0.4	-812	-32479	2000	16.2	32479	32479	0
-0.8	-1583	-33250	2100	15.8	33250	33250	0
-1.2	-2317	-33984	2200	15.4	33984	33984	0
-1.6	-3016	-34683	2300	15.1	34683	34683	0
-1.9	-3682	-35349	2400	14.7	35349	35349	0
-2.3	-4318	-35985	2500	14.4	35985	35985	0
-2.6	-4926	-36593	2600	14.1	36593	36593	0
-2.9	-5507	-37174	2700	13.8	37174	37174	0
-3.2	-6064	-37731	2800	13.5	37731	37731	0
-3.5	-6597	-38264	2900	13.2	38264	38264	0
-3.7	-7109	-38776	3000	12.9	38776	38776	0

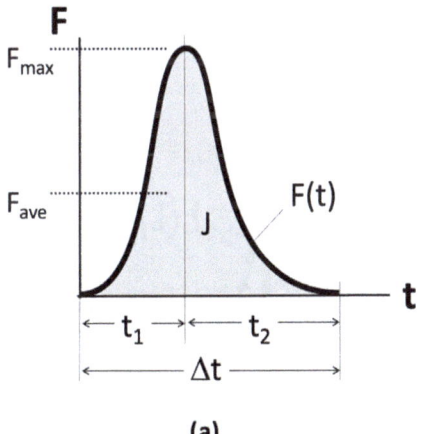

(a)

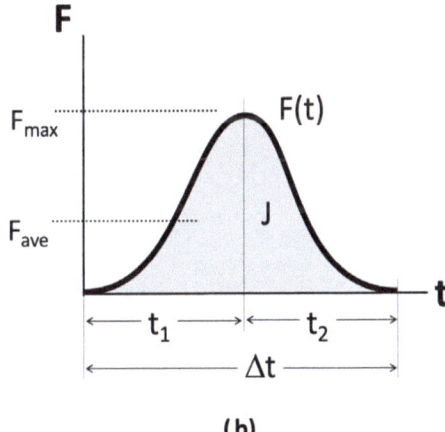

(b)

Fig. 3.29 Force response profile examples

At the beginning of a collision event where one object is initially stationary, the velocity difference between the two objects is high, and therefore the applied force can also be high, per Newtons second law…

$$F = ma \tag{3.35}$$

$$F = m\Delta v \tag{3.36}$$

The velocity difference between the two objects reduces as the impulse event progresses and momentum is transferred. This phenomenon explains the behavior of the F(t) profile in the t_2 timeframe.

The presence of **friction**, and one that resists motion throughout the event, will increase the F(t) and can prolong the duration of the impulse event, as illustrated by F_f(t) in Fig. 3.30.

The performance development for some crashworthiness loadcases involves tuning structure such that the impulse time duration is strategically manipulated in order to moderate the collision force. Understanding the relationship between the impulse time duration, the collision force and; the test vehicle's mass, the stiffness of its structure, the strength of its structure, and the effect of significant frictional force(s) are important for successful, efficient, structural design.

Stored Energy and its Release in Plastic Collision Objects

Energy will be stored in two zones of a plastic collision object during a collision event; in the unyielding back-up structure and within the crushed structure. The back-up structure

Fig. 3.30 Effect of friction on the force response profile

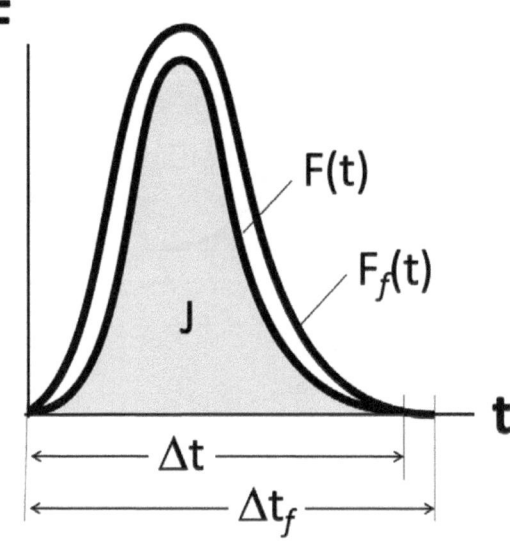

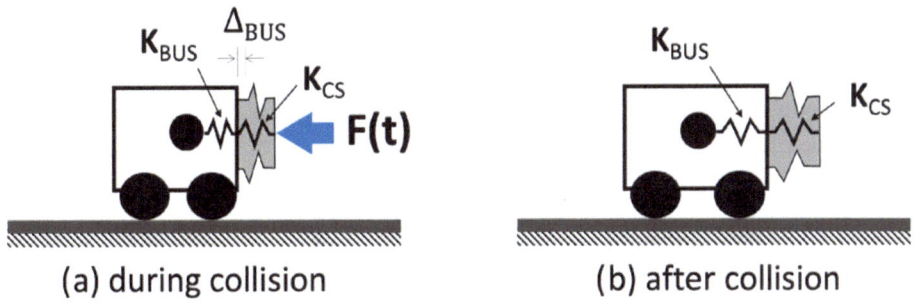

(a) during collision   (b) after collision

Fig. 3.31 FDB illustrating stored energy mechanism

being defined as the structure between the applied impulse force and the object's generalized mass center. The amount of energy stored will be dependent on the respective spring rates, K_{BUS} and K_{CS}, and the magnitude of elastic compression in each zone (Δ_x), as illustrated in Fig. 3.31. Although the magnitude is often very small, the back-up structure and the crushed structure of a collision object will elastically compress during the collision. This stored energy will release as the impulse force is removed. The unyielding back-up structure will return to its original size while the crushed structure will expand slightly.

The release of this stored energy will cause a collision object to 'spring-back' from the other collision object; behavior often visible in crash videos.

The amount of energy stored as elastic compression can be found by using Hooke's Law for Potential Energy, which states that the stored energy is a function of the spring rate and the elastic displacement…

$$E_{stored} = \frac{1}{2}K\Delta^2 \tag{3.37}$$

$$E_{stored} = \frac{1}{2}K_{BUS}\Delta^2_{BUS} + \frac{1}{2}K_{CS}\Delta^2_{CSe} \tag{3.38}$$

where:

E_{stored} = energy stored within a collision object

K_{BUS} = stiffness of a collision object's back-up structure

Δ_{BUS} = maximum elastic compression seen within a collision object's back-up structure

K_{CS} = stiffness of a collision object's crushed structure

Δ_{CSe} = maximum elastic compression seen within a collision object's crushed structure

Note that Eq. 3.38 is developed simply to reinforce a mental model of the behavior and relationship to the influencing parameters. The practical usage of the equation in the vehicle development process is limited.

In actuality, compression within the back-up structure will occur between the applied impulse force and the mass elements of the vehicle, as the masses are the elements which are 'resisting momentum change'. For an example, consider an automobile as a collision object and its back-up structure between the impulse force and the heavy liquid fuel tank (Fig. 3.32). Imagine this structure as a spring and how it will compress during the event, even if the magnitude of compression is very small. For a complete representation of this vehicle's K_{BUS}, consider that there is a spring between the impulse force and every ungrounded mass element of the vehicle.

With this understanding, you can see that determining an accurate value for K_{BUS} would be quite complex, especially if you consider that there might be more than one location of the applied impulse force and that the location(s) might change throughout the event.

In subsequent chapters, the stiffness of a non-stationary collision object is simplified as the stiffness between the applied impulse force and the generalized mass center of the object; thus assuming this stiffness approximates the compilation of the springs associated with each mass element within the vehicle and the crushed structure (K_{CS}). Variables K_{Sm} and K_{Tm} are used to represent a generalized stiffness for the striking object and target object respectively (thus the subscripts "S" and "T"). The subscript "m" signifies that the spring rate is between the applied force and the object's generalized mass center.

Stored Energy and its Release in 'Rigid' Collision Objects
It is important to remember, no matter what adjective is used to describe a barrier or structure, nothing is truly rigid. The condition where a collision object strikes a 'rigid', unyielding barrier fixed to the ground, an example shown in Fig. 3.2, is no exception. In this collision, the impulse force is equal and opposite, applying force to the striking object

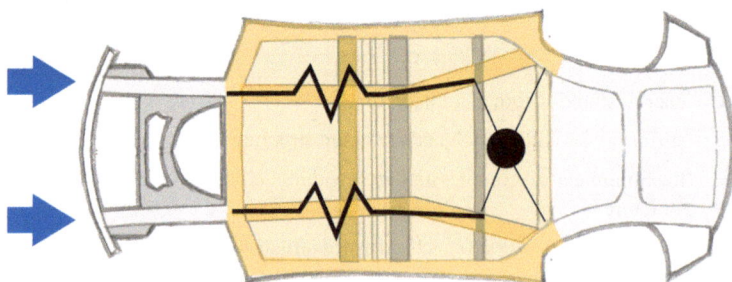

Fig. 3.32 Back-up structure spring visualization, liquid fuel tank

and to the stationary barrier. The displacement of the stationary barrier can be determined using Hooke's Law of Elasticity…

$$F = K\triangle \tag{3.39}$$

Equation 3.39 tells us that, for a given force, the displacement reduces as the stiffness increases, which explains the perception that exceptionally stiff objects are rigid.

Energy stored within exceptionally stiff, unyielding objects follows Eq. 3.37, however you'll note that the displacement is more influential than the stiffness when it comes to the magnitude of stored energy.

3.3   Occupant Behavior and Considerations

The Other Collision

Although occupant restraint systems are outside the scope of vehicle structure, it's important to recognize that restraint systems work in conjunction with the vehicle structure to provide occupant performance during a crash. An understanding of restraint system function and behavior is a required for those involved in automotive structural development. Considering this, it's appropriate to dig deeper into the physics and kinematics of occupant protection.

Consider the front crash vehicle/occupant FBD shown in Fig. 3.33, which illustrates that there are a limited number of connections between the occupant and the vehicle; only the seat belt attachment locations. Considering this, and the fact that the human body is quite flexible, much of the occupant's body will remain in motion during a crash event even though the vehicle is slowing down.

Fig. 3.33 Vehicle/occupant FBD

Thus, the deceleration characteristic of the structure alone is not enough to ensure occupant safety. The restraint system, including the seat belts, the airbags, elements of the instrument panel (to manage knee forces in frontal loadcases) and door trim in side impact loadcases, provide an accompanying function to complete the job. The interaction between the occupant and the restraint system is referred to here as 'the other collision'. The goal of the vehicle structure is to provide deceleration characteristics such that the restraint system can successfully manage this 'other collision'.

Collision Parameters from the Occupant's Perspective

The impulse Eq. 3.7 duplicated here below (as Eq. 3.40) tells us that there is a realation-ship between the force magnitude applied to a collsion object and the time duration of the impulse event; as the time duration reduces, the force increases.

$$J = F_{ave}\Delta t \tag{3.40}$$

Similarly, the energy Eq. 3.29 duplicated here below (as Eq. 3.41) tells us that there is a relationship between the force magnitude and the crush distance; as the crush distance reduces, the force increases.

$$F_{ave}\Delta = \Delta KE \tag{3.41}$$

Finally, Newton's second law shows a relationship between the force magnitude and acceleration; an object that decelerates quicker (or slows down over a shorter amount of time) will have higher forces exherted onto it.

$$F = ma \tag{3.42}$$

$$F = m\left(\frac{v_o - v_f}{\Delta t}\right) \tag{3.43}$$

To help rationalize these relationships, consider the 'jumping into bed' analogy illus-trated in Fig. 3.34. Jumping into a soft bed will produce a collision that occurs over a longer period of time and with a larger maximum displacement than when jumping into a slab of bricks. The 'collision' with the soft bed will result in a smaller maximum acceleration and a smaller maximum force.

Another fundamental of minimizing the force in a collision is to maximize the area over which it is distributed. Consider the difference between jumping into a soft bed in the two body positions illustrated in Fig. 3.35; distributing the collision force will result in a smaller displacement and a smaller maximum force.

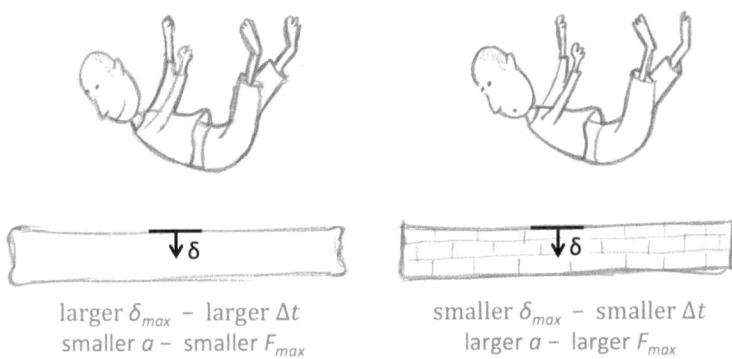

Fig. 3.34 Analogy for Newtons second law

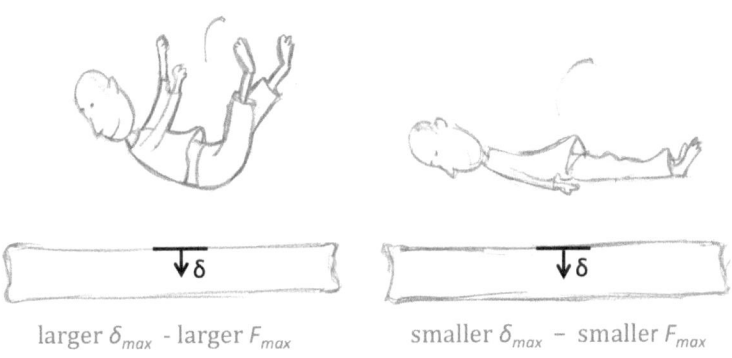

Fig. 3.35 Analogy for force distribution

Characteristics of the 'Other Collision'

The function of the restraint system is to manage the deceleration of the occupant such that the maximum deceleration magnitude and maximum force (as witnessed by the occupant) are kept to a manageable level.

For example, consider a frontal crash event. Although the structure's behavior dictates the deceleration characteristics of the seat belt attachment locations, the restraint system can function to moderate occupant deceleration. Figure 3.36 illustrates the behavior of the seat belt and airbag. Once a crash is sensed, the seatbelt tensions to pull the occupant towards the seat, which will increase the distance over which the occupant can be decelerated, as illustrated in Fig. 3.36b. The belt system can then release belt length to increase the 'collision' time, as illustrated in Fig. 3.36c. The airbag finishes the occupant deceleration event, as illustrated in Fig. 3.36d.

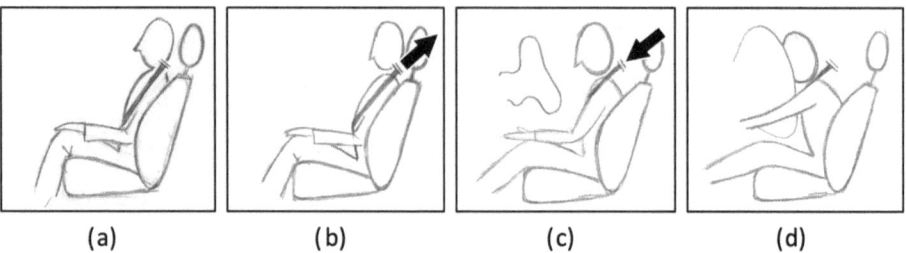

| (a) | (b) | (c) | (d) |

Fig. 3.36 Timesteps of the occupant in a frontal collision

This behavior of the seat belt and airbag systems can be witnessed in the reference video, Link-3.3.

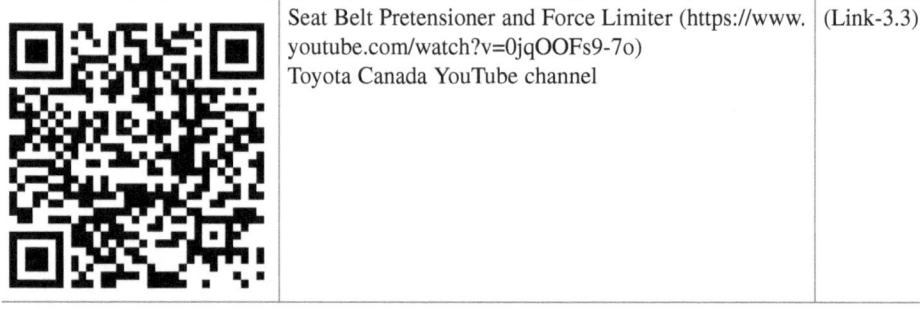

| | Seat Belt Pretensioner and Force Limiter (https://www. youtube.com/watch?v=0jqOOFs9-7o) Toyota Canada YouTube channel | (Link-3.3) |

A similar story plays out in side impact crash events, however the ability of the seat belt to control occupant kinematics is greatly reduced. In side impact, occupant deceleration is controlled by the *roof-rail-airbag* and the *side door trim*. Perhaps not surprisingly, the door trim can be strategically shaped such that collision forces are not overly concentrated and energy absorption foam is typically found between the door trim and the door structure to moderate the impulse time duration.

Collision Loadcases

4

Abstract

An auto manufacturer considers many collision loadcases when developing a new automobile or enhancing the performance of an existing one. These loadcases have been developed in an attempt to represent the worst-case loading conditions that typically occur in the field. Many of these loadcases can be considered "driving loadcases"; ones which tend to define portions of the vehicle structural loadpath topology or the strength or stiffness levels within it. Crashworthiness loadcases that typically drive the vehicle structural loadpaths are covered in detail within this chapter, including discussion of how the EV configuration affects the structure and how the structure might change to accommodate the demands of an EV. The material in this chapter is organized and presented with the goal of building the reader's intuition of structural behavior in mind.

Particularities of EVs

The introduction of high voltage electronics into the automobile has required some additional safety considerations to be applied to many standardized loadcases, including many of the driving loadcases covered in this chapter.

The US NHTSA established explicit requirements for Side-MDB, Rear Barrier, 0° Full-Frontal, and 30° Full-Frontal. FMVSS305 documents these additional requirements, which include;

Supplementary Information The online version contains supplementary material available at https://doi.org/10.1007/978-3-031-75933-8_4.

- a specified maximum electrolyte spillage
- no evidence of electrolyte leakage into the passenger compartment
- all components of the energy storage unit must be attached to the vehicle post-test
- stipulations on direct access to high voltage content

The EU established similar requirements for EVs with the introduction of UNIECE. This regulation specifies post-test requirements for 0° Full-Frontal, MOF, Side MDB, and Rear Barrier. Requirements include;

- protection from electrical shock
- stipulations on electrolyte spillage
- retention of the propulsion battery housing

4.1   Full-Frontal

"Full-Frontal" is the term that the automotive industry typically uses to describe the "Zero Degree Rigid Wall/ Full-Frontal" crash test. This loadcase is a regulatory crashworthiness loadcase in many parts of the world, FMVSS208 in the US, UNR137 in the EU, and GB11551-2014 in China, for example. It can also be found in NCAP assessments as well.

In these loadcases, a vehicle traveling at a prescribed speed strikes a fixed barrier. The entire width of the vehicle contacts the barrier (thus the use of the descriptor "full") and the barrier is orientated perpendicular to the vehicle's direction of travel (this reference angle is referred to "zero degree" within the industry); Figs. 4.1 and 4.2 illustrate.

The following videos (Link 4.1, Link 4.2) show the Full-Frontal test condition.

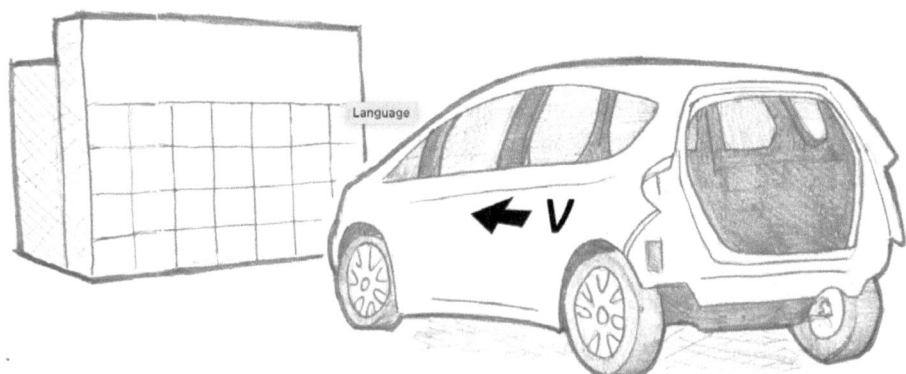

Fig. 4.1 General full-frontal test condition

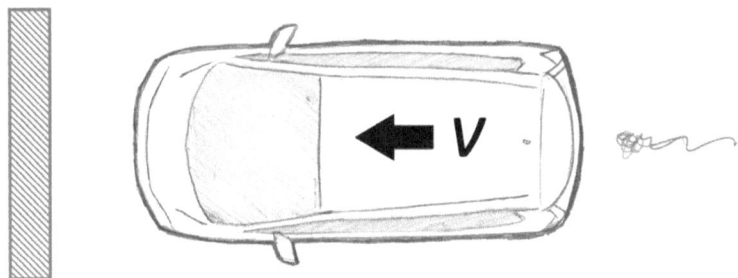

Fig. 4.2 Full-frontal barrier overlap

	Full Frontal test of a 2009 Honda Fit NHTSA, USNCAPn	(Link-4.1)

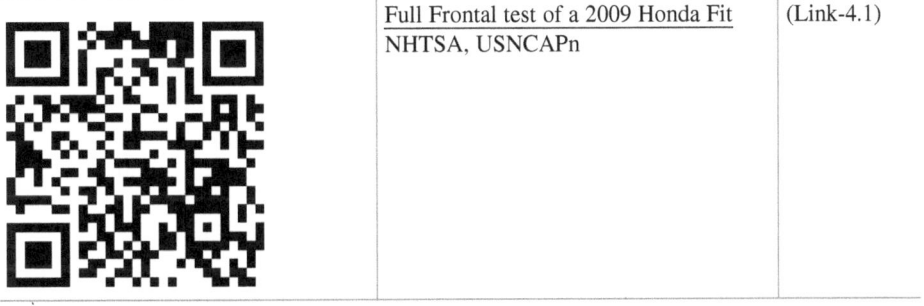	Full Frontal test of a 2020 Volvo XC60, slow motion NHTSA, USNCAP	(Link-4.2)

Performance in this loadcase is based on occupant performance metrics extracted from the crash dummies used in the test.

Notice how a few non-structural components have been removed from the vehicle in Fig. 4.1 and the Honda Fit example video. Because mass is recognized as something that will affect crashworthiness performance, each test protocol will stipulate the test mass of a vehicle; usually based on its curb weight. Removing non-structural components is done to offset the weight of on-board measurement equipment and enable the test vehicle to achieve the test weight specified in the protocol.

4.1.1 Loadcase Modeling

The test vehicle approaches and strikes a stationary 'rigid' barrier in this loadcase. We can utilize the collision model developed in Sect. 3.2.3.5 by considering that the target object is essentially the barrier and the earth that it is attached to; $m_T \gg m_S$. We would therefore expect unmeasurable velocity change of the barrier/earth and a significant amount of velocity change in the striking vehicle; certainly consistent with what is observed in physical testing. Since the barrier has no plasticity and spans the entire width of the test vehicle, the test vehicle's final velocity is zero and all its initial energy is expended as material failure and crush. The collision model is illustrated here in Fig. 4.3.

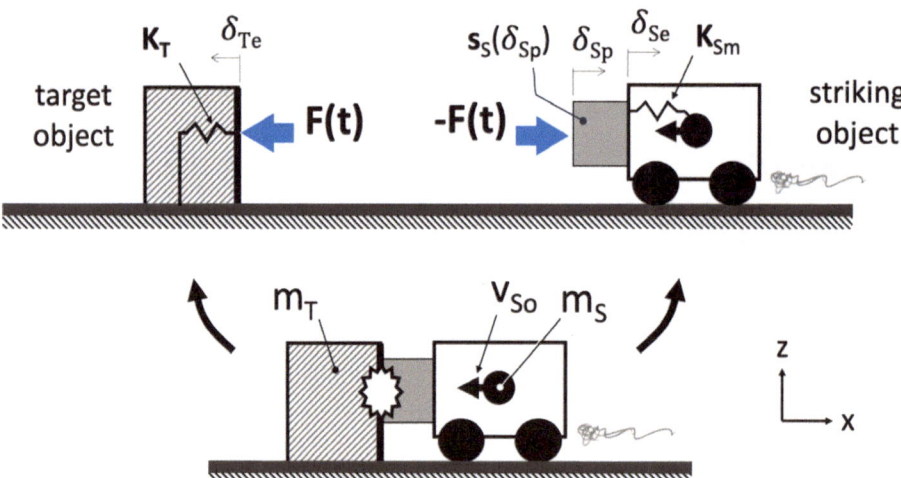

Fig. 4.3 General full-frontal collision model

Fig. 4.3 variable definitions:

s_S	=	is the strength profile of the striking vehicle,
δ_{Sp}	=	is the direction of striking vehicle plastic deformation measure,
δ_{Se}	=	is the direction of striking vehicle elastic deflection measure,
K_{Sm}	=	is the stiffness of the striking vehicle,
δ_{Te}	=	is the direction of barrier elastic deflection measure,
K_T	=	is the stiffness of the barrier

- Plasticity (s_S): Plasticity exists only in the striking vehicle, assuming design and construction of the barrier is adequate.
- Elasticity (K_{Sm}, K_T): Elasticity exists in both the striking vehicle and barrier. The relevant stiffness of the striking vehicle is between the applied impulse force and its generalized center of mass. If the barrier has been designed and constructed well, $K_T \gg K_{Sm}$ and deflection within the barrier will be extremely small.

Insight from Governing Equations

The impulse equation can be leveraged from our collision model. Here, impulse is defined as the change in a collision object's momentum (Eq. 4.1), but also equivalent to the area under the F(t) curve (Eq. 4.2).

$$\text{Impulse}: J = p_o - p_f \qquad [\text{N s] or [kg m/s}] \tag{4.1}$$

$$\text{Impulse}: J = \int_o^f F(t)dt \tag{4.2}$$

or

$$J = F_{ave}\Delta t \tag{4.3}$$

where:

J	=	impulse
p	=	momentum of one collision object
F	=	iimpulse force
Δt	=	time duration of collision

...equating these two...

$$J = J \tag{4.4}$$

$$F_{ave}\Delta t = p_o - p_f \tag{4.5}$$

$$F_{ave}\Delta t = m_S v_{So} - m_S v_{Sf} \tag{4.6}$$

$$F_{ave}\Delta t = m_S v_{So} - m_S(0) \tag{4.7}$$

$$\boxed{F_{ave}\Delta t = m_S v_{So}} \tag{4.8}$$

where:

p_o	=	momentum of the striking vehicle, just prior to the collision
p_f	=	momentum of the striking vehicle, after the collision
m_S	=	mass of the striking vehicle
v_{So}	=	initial velocity of the striking vehicle
v_{Sf}	=	final velocity of the striking vehicle
F_{ave}	=	average impulse force
Δt	=	time duration of collision

Equation 4.8 shows us that the F·t product required to slow the striking object is dependent on its mass. This should feel intuitive, as a heavier vehicle will have more initial kinetic energy and slowing it to a stop will require for effort. We will also find that a heavier vehicle will need to absorb more energy than its lighter counterpart.

From a structural engineering standpoint, we are particularly interested in how the vehicle's kinetic energy is dissipated during the event.

From the conservation of energy law...

$$E_f = E_o \tag{4.9}$$

$$(KE_f + W) = KE_o \tag{4.10}$$

$$(0 + W) = \left(\frac{1}{2}m_S v_{So}^2\right) \tag{4.11}$$

$$W = \frac{1}{2}m_S v_{So}^2 \tag{4.12}$$

$$\boxed{F_{ave}\Delta_{Sp} = \frac{1}{2}m_S v_{So}^2} \tag{4.13}$$

where:

m_S	=	mass of the striking vehicle
v_{So}	=	initial velocity of the striking vehicle
v_{Sf}	=	final velocity of the striking vehicle
$Fave$	=	average impulse force
Δ_{Sp}	=	crush distance seen within the striking vehicle

   ... we find that the striking vehicle's kinetic energy will be converted into material fracture and crush within the test vehicle. Equation 4.13 also tells us that there is a relationship between the average impulse force and the magnitude of crush seen within the striking vehicle; as the amount of total crush increases, the average impulse force decreases. Consider the 'jumping into bed' analogy illustrated in Fig. 3.34 to solidify your understanding.

System Energy Flow
Flow of system energy during the collision is rather simple, thanks to the unyielding barrier. As Fig. 4.4 illustrates, all the initial kinetic energy is converted into crush in the striking vehicle ($F_{ave}\Delta_{Sp}$). Some energy is stored elastically during the event, the amount of which is dependent on the barrier and striking vehicle stiffnesses (K_T and K_{Sm}, respectively) and the maximum elastic compression occurring within these two objects (Δ_{Te} and Δ_{Se}, respectively). This energy is released after the impulse force is removed and 'springs' the striking vehicle off the barrier.

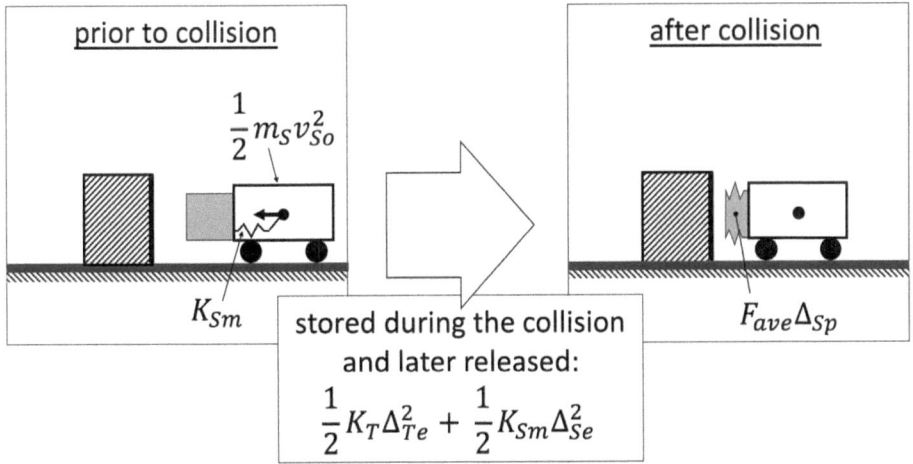

Fig. 4.4 Full-frontal energy conservation

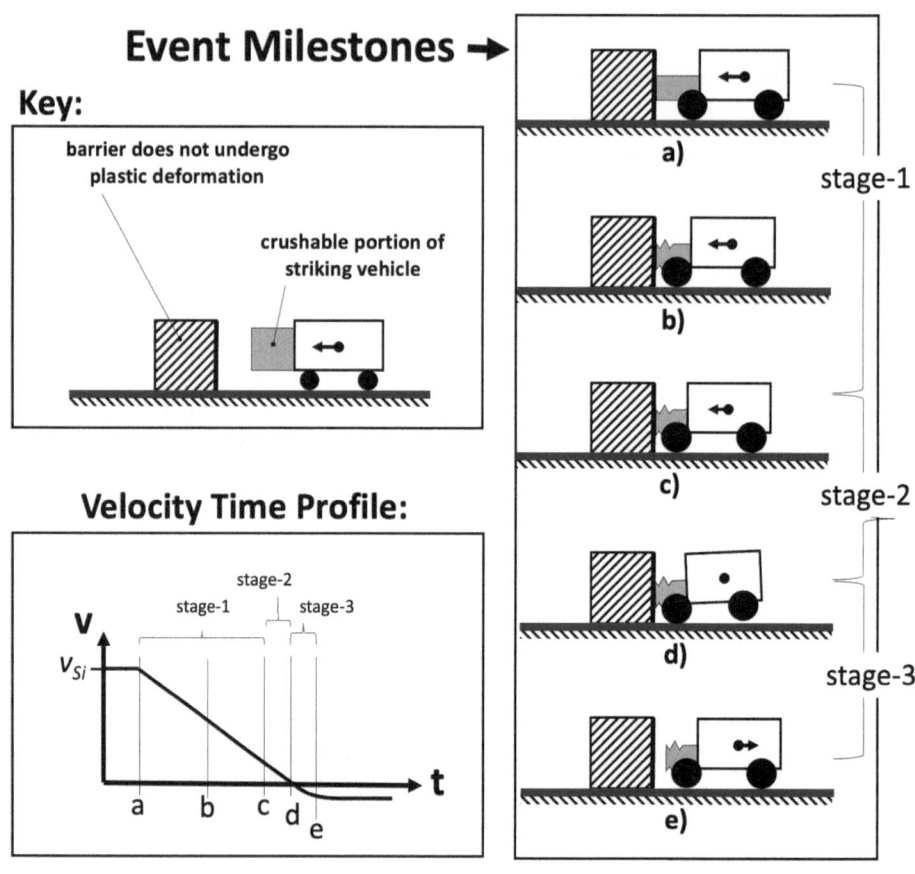

Fig. 4.5 Full-frontal event milestones

4.1.2   Fundamental Behavior and Physics

Event Milestones
The time steps of the Full-Frontal collision are shown in Fig. 4.5. We can subdivide this loadcase, or event, into two distinct stages: crush and elastic deformation.

Behavior, High-Level
Stage-1: Crush and Elastic Compression

(a) The striking vehicle first contacts the barrier.
(b) Crush occurs in the striking vehicle and its velocity reduces as energy is dissipated. The striking vehicle absorbs energy through elastic compression.

Stage-2: Only Elastic Compression

(c) The impulse force is no longer high enough to cause material fracture and crush however the target vehicle still has velocity. During this short duration of time, the striking vehicle compresses further. A vertical misalignment between the vehicle's cg and the impulse force will instigate a side-view rotation (pitch).

Stage-3: Rebound

(d) The striking vehicle has stopped moving forward. The vehicle and barrier have stored elastic energy.
(e) The energy stored is released and the vehicle 'springs off' of the barrier.

Behavior, Detailed

Stage-1: Crush and Elastic Compression

This stage begins with the vehicle's contact with the barrier and is characterized by the crush of both the structure and other elements of the vehicle. The most significant crush within the structure occurs in portions of the *body front-compartment-mid-rail* and the *front-cradle*. The front crash strategy also involves deformation of the *front-cradle-rails*.

Some elements of the vehicle are intentionally designed and tuned to perform a significant energy absorption role in this event. These include; the *impact-beam,* foam on the front of the impact beam, the forward portion of the *front-compartment-mid-rails*, and portions of the *front-cradle* structure.

Energy absorption is not a significant design parameter of other elements crushed during the event. These components have varying levels of contribution to energy reduction. These elements include; the *radiator*, *hood*, and *headlamps*.

The vehicle structure can be subdivided into segments based on their participation in the crush stage, as illustrated in Fig. 4.6.

- *crush zone*: where crush is to occur
- *back-up structure*: the rearward area that remains unyielding during the event
- *transition zone*: where lower levels of crush might occur.

The force at which the *rail* in the crush zone begins to deform is called its "load capacity". To ensure that *rail* deformation occurs in the crush zone, the *rail* in the transition zone and back-up structure must be able to manage the load capacity force without deforming; thus, the rail strength in the transition zone and back-up structure must higher than that in the crush zone; $S_{BUS} > S_{TZ} > S_{CZ}$, as illustrated in Fig. 4.7.

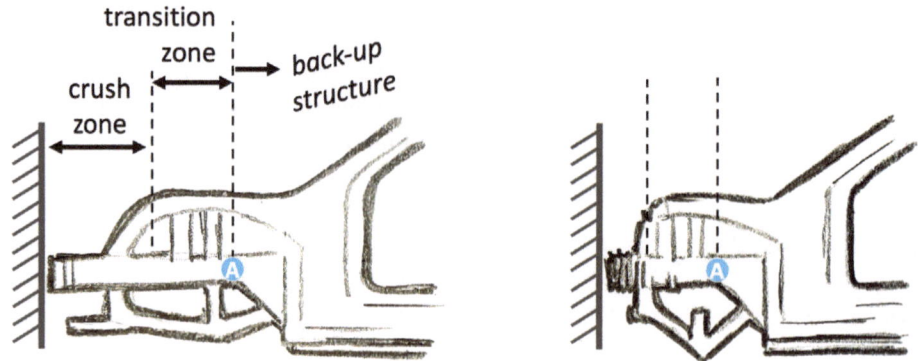

Fig. 4.6 Zones of the front structure

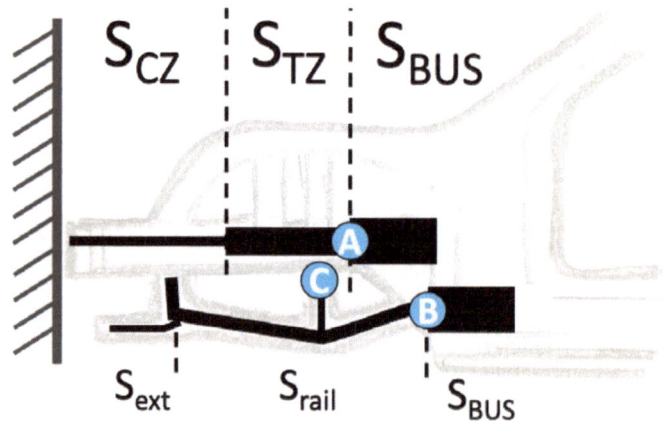

Fig. 4.7 Strength profile of the front structure

A similar story plays out with the cradle structure. The cradle rear attachment location on the body structure (B in Fig. 4.7) must be stronger than the *cradle-side-rail* (S_{rail}), and the *cradle-side-rail* must be stronger than the crushable *cradle-rail-extension* (S_{ext}). Furthermore, in vehicles that have a cradle mid-mount and a front crash strategy that includes a V-bend cradle deformation mode, the cradle mid-mount (C in Fig. 4.7) must be tuned such that it is strong enough for vehicle usage loadcase but not so strong that it disallows the cradle V-bend (downward buckling of the *cradle-side-rail* between its forward-most attachment to the body structure and location B).

Progressive crush and a V-bend deformation mode of the *front-cradle* can be seen in the following example video (Link 4.3); with regards to progressive crush, note the crush of the *front-cradle-rail-extensions* prior to the V-bend deformation of the *cradle-side-rails*.

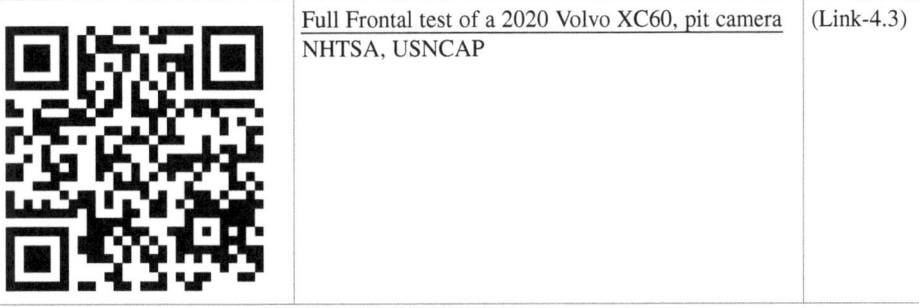

	Full Frontal test of a 2020 Volvo XC60, pit camera	(Link-4.3)
	NHTSA, USNCAP	

As strength progressively increases within the vehicle, we can imagine a generalized impulse force profile like shown in Fig. 4.8; where the impulse force increases as the loadpath strength increases.

Equation 4.13 illustrates the importance of crush; it reduces the force applied to the structure. Crush also increases the time that the vehicle and vehicle occupants slow down; thus reducing the acceleration and force applied to the occupant due to that deceleration. A few relationships between crush and the structural implications or performance can be derived; three follow here…

(1) The relationship between crush and rail force can be shown by using the conservation of energy Eq. 4.13 developed above and duplicated here.

$$F_{ave} \Delta_{Sp} = \frac{1}{2} m_S v_{So}^2 \qquad (4.14)$$

Fig. 4.8 Generalized full-frontal impulse force profile

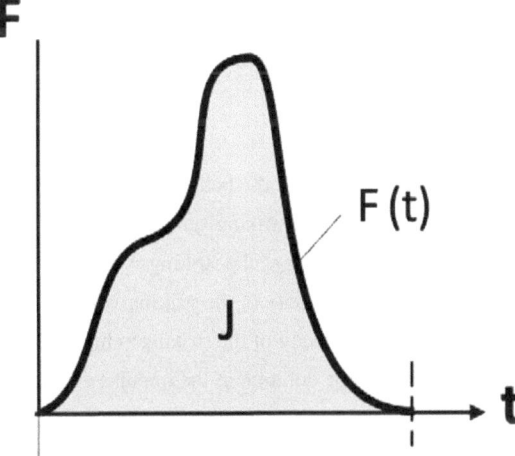

$$F_{ave} = \frac{\frac{1}{2}m_S v_{So}^2}{\Delta_{Sp}}$$

(4.15)

where:

m_S =   mass of the striking vehicle

v_{So} =   initial velocity of the striking vehicle

F_{ave} =   average impulse force

Δ_{Sp} =   crush distance seen within striking vehicle

Equation 4.15 can be used to estimate the force magnitude seen by the *front-compartment-mid-rail* given different crush distances, or vice versa. Understand that it is certainly an estimation, as it assumes only the *mid-rail* performs energy absorption. Notice the relationship between crush space and the force applied to the structure; as crush distance increases, the force decreases.

(2) The relationship between the generalized force applied to the *rail* and the time duration of crush can be developed by exercising Newton's second law of motion equation.

$$F_{rail} \approx m_S a_S$$

(4.16)

$$a = \frac{\Delta v}{\Delta t}$$

(4.17)

$$F_{rail} \approx m_s \frac{v_{so} - v_{sf}}{\Delta t} \quad v_f = 0$$

(4.18)

$$F_{rail} \approx m_s \frac{v_{so}}{\Delta t}$$

(4.19)

where:

F_{rail} =   force seen by the body mid-rail

m_S =   mass of the striking vehicle

a_S =   deceleration of the striking vehicle

v_{So} =   initial velocity of the striking vehicle

v_{Sf} =   final velocity of the striking vehicle

Δt =   the time duration of the impulse event

Note the relationship between the duration of deceleration and the generalized *rail* force at the front of the back-up structure shown in Eq. 4.19; as the deceleration time increases, so does the force seen by the *rail*. Again, understand that it is an approximation, as it assumes that only the body *mid-rails* participate in managing the loads. From this relationship and the one prior, it is evident that engineering a back-up structure to be unyielding becomes more difficult as crush is reduced.

(3) Working further, we can use Eq. 4.13 (duplicated here as Eq. 4.20) and Newton's second law of motion understand the relationship between crush space and the vehicle's deceleration rate:

$$F = \frac{\frac{1}{2}m_S v_o^2}{\Delta_{Sp}} \tag{4.20}$$

$$F = ma \tag{4.21}$$

where:

a	=	deceleration rate of the striking vehicle
m_S	=	mass of the striking vehicle
v_o	=	initial velocity of the striking vehicle
Δ_{Sp}	=	crush distance seen within striking vehicle

$$F = F \tag{4.22}$$

$$\frac{\frac{1}{2}m_S v_o^2}{\Delta_{Sp}} = m_S\, a \tag{4.23}$$

$$a = \frac{v_o^2}{2\Delta_{Sp}} \tag{4.24}$$

We can use Eq. 4.24 to estimate the generalized deceleration rate of the vehicle. The deceleration rate of the vehicle is of interest because the likelihood of occupant injury is related to the occupant deceleration rate and the occupant deceleration rate is significantly influenced by the vehicle deceleration; recall the lessons of Sect. 3.3, "The Other Collision".

Manufacturers typically allocate around 600 mm of front crush space, or 'free crush space', to achieve a deceleration that is conducive to restraint system development and ultimately occupant performance.

Keeping occupant deceleration within reasonable levels, based on the capability of the human body to manage acceleration/ deceleration, is a critical function of structural performance in the Full-Frontal loadcase. Ultimately, the structure works in conjunction with the occupant restraint systems to achieve occupant safety performance, as described in further detail in Sect. 3.3, "The Other Collision".

Grounded Components

There are components carried by the structure that will contact the barrier directly during the crush stage; the wheel, for example. Prior to this contact, there is a ride-down loadpath in the structure between the barrier and the component; recall the front crash loadpath derived in Sect. 1.4.2. Once a component contacts the barrier, the structure no longer needs to manage the force associated with the component's deceleration ($m_{component}a$) and the energy associated with the component ($\frac{1}{2} m_{component} v^2$) is immediately deducted from the vehicle's energy. Figure 4.9 illustrates select components, their individual energies, and the force that they apply on the structure during the Full-Frontal loadcase.

Typically, the front wheel is a component that 'grounds-out' during a Full-Frontal test. When the wheel contacts the barrier, the forces associated with the wheel and brake deceleration no longer need to be managed by the structure. These components have 'bottomed-out' on the barrier; their energy has been 'grounded'. The striking vehicle energy profile could be illustrated by the profile shown in Fig. 4.10, where abrupt energy reductions are due to heavy components bottoming out.

In many ICE automobiles, the front powertrain is another component that grounds-out during the Full-Frontal test. It can be seen in the following video of a 2020 Volvo XC60., Link 4.4

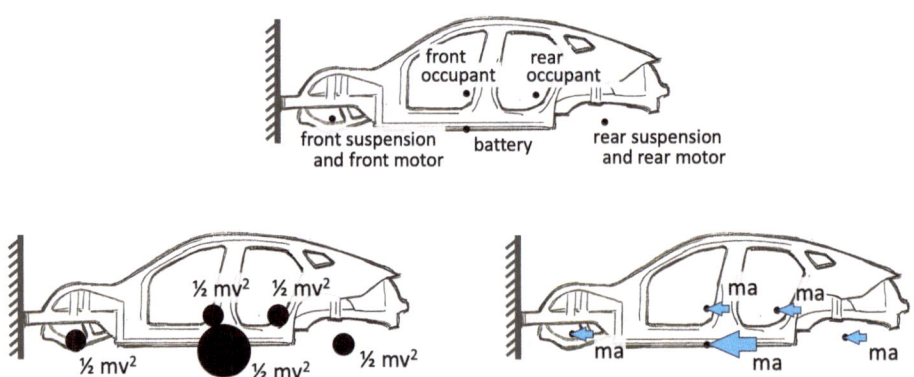

Fig. 4.9 Select components, their energies, and forces

Fig. 4.10 Effect of grounded
components on striking vehicle
energy

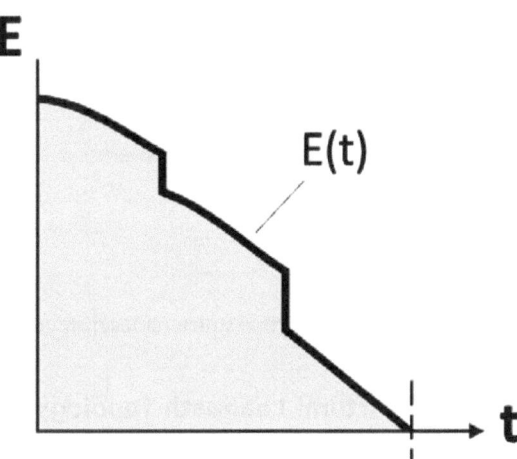

	Full Frontal test of a 2020 Volvo XC60, pit camera NHTSA, USNCAP	(Link-4.4)

Stage-2: Only Elastic Compression

The second stage of the event is characterized by the elastic deformation of the vehicle, primarily the vehicle structure. In this stage, crush of the structure no longer occurs and the remaining vehicle energy is expended in elastic deformation of the back-up structure and crushed structure, as detailed in Sect. 3.2.3.

Depending on the height position of the cg relative to the structure's point of contact on the barrier, the vehicle might exhibit a pitch motion during this stage.

Stage-3: Rebound

Removal of the impulse force releases the stored energy stored in the barrier and striking vehicle resulting in the striking vehicle 'springing off' of the barrier.

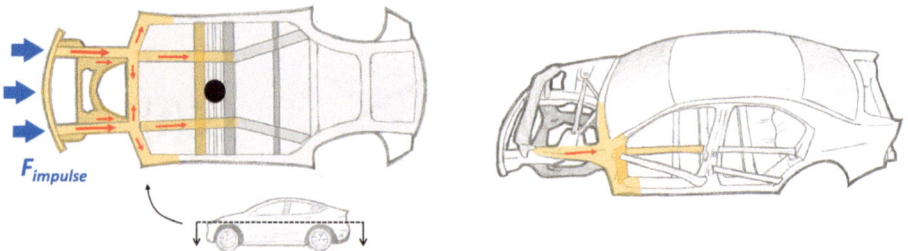

Fig. 4.11 Typical full-frontal structural loadpath topology

4.1.3   Structural Loadpath Topology

The Crush strategy portion of the Full-Frontal loadpath topology includes all the energy absorption elements on the front-compartment, including the *front-impact-beam*, *front-compartment-mid-rails*, and *front-cradle*. The energy ride-down portion of this loadcase's loadpath topology, the portion managing strength and stiffness between the barrier and the vehicle's mass concentrations, is largely comprised of the *dash* area structure, *center-compartment-longitudinal-rails*, and *rockers*. Contribution of further rearward structure is possible, depending on the vehicle's mass distribution.

Some manufacturers consider the front-side door structure as an important part of the Full-Frontal loadpath topology and engineer the beam(s) accordingly. Figure 4.11 illustrates the typical loadpath topology for Full-Frontal.

Because Full-Frontal engages both *front-compartment-mid-rails* and only one is engaged in the MOF loadcase, it can be challenging to engineer the *rail* force capacity and crush behavior such that it performs well in both conditions.

Geometric initiators, as illustrated in Fig. 4.12, are design elements often employed to 'tune' the crush behavior of components within the crush zone the body *front-compartment-mid-rails* and the *cradle-side-rails*, in particular.

4.1.4   Peculiarities of EVs

Post-Crash High Voltage Safety
Recall the overview of FMVSS305 and UNIECE regulations provided in this chapter's introduction. These regulations apply to vehicles with high voltage content and specify post-crash performance related to; protection from electrical shock, level of electrolyte spillage, and retention of the propulsion battery housing. Both regulations apply to the Full-Frontal loadcase.

Thoughtful packaging of high voltage components and wiring with respect to crush zones is required to efficiently design for these regulations and any other high voltage

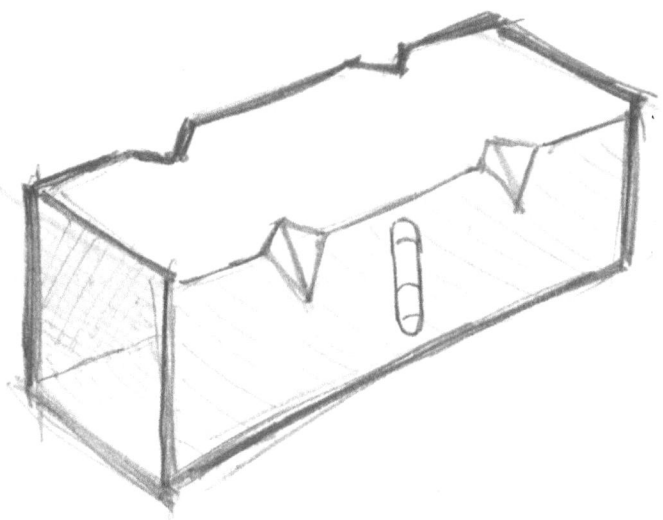

Fig. 4.12 Example of initiation features

safety requirements that individual manufacturers maintain. Shielding can be incorporated to protect high-voltage content packaged within a crush zone, however these solutions tend to add cost and mass to the automobile. Damage to high voltage content can also occur due to low displacement elastic deflections within the structure, particularly content within the propulsion battery housing. Thus, it might be appropriate to investigate the magnitude of static displacement that occurs in Full-Frontal near sensitive components, depending on cell chemistry and the electrical architecture within the housing.

Structural Topology

Recall that Chap. 2 described that, in the era of 600 Wh/l battery cell technology and vehicle range requirement of 300 miles or more, the volume of battery cells required dictates that the *center-compartment-mid-rails* are eliminated from the body structural topology, in an underfloor BEV configuration (Fig. 2.3). Considering that adequate fore/aft body structure strength at the root of the front-compartment-mid-rail (A in Fig. 4.13) is a critical for front crashworthiness performance, the loss of this loadpath requires some local reengineering.

The body structure will have to be strengthened or its loadpaths modified unless the propulsion battery housing structure is engineered to have meaningful contribution to the back-up structure. Figure 4.14 illustrates some methods of strengthening existing loadpaths. Loadpaths can be created or enhanced between the root of the *front-compartment-mid rail* and *the A-pillar* and *tunnel*, as seem in many Geely/Volvo

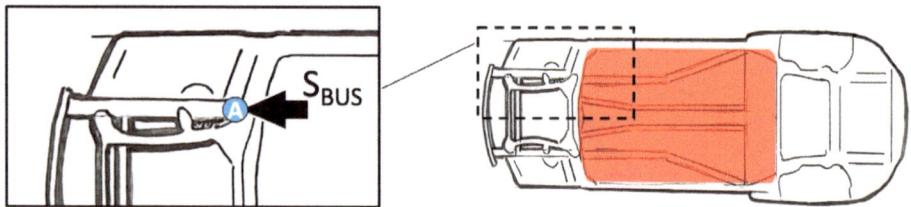

Fig. 4.13 Location of required strength for the full-frontal loadcase

automobiles. The body *front-toque-box* chamfer can be increased and the cradle structure extended, as seen in the 2021 Ford Mustang Mach-E. Note that some solutions will
impact the propulsion battery volume.

Figure 4.15 illustrates methods to strengthen the front crash back-up structure through
the use of the propulsion battery housing structure. The front of the propulsion battery
housing side-rail can be deliberately orientated to back-up the cradle rear attachment, as
executed in the 2018 Jaguar I-Pace. Longitudinal structure within the propulsion battery
housing can be employed to supplement the body structure's back-up structure, as with the
2018 Tesla Model-3. Once again, that some solutions will impact the propulsion battery
volume.

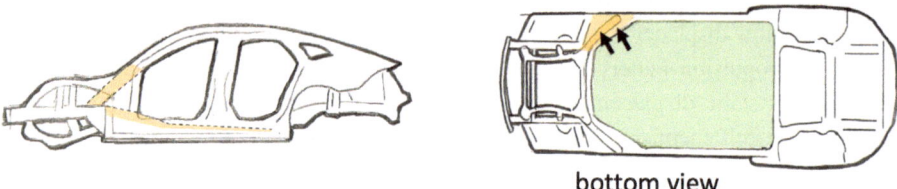

Fig. 4.14 Topology design elements, side view

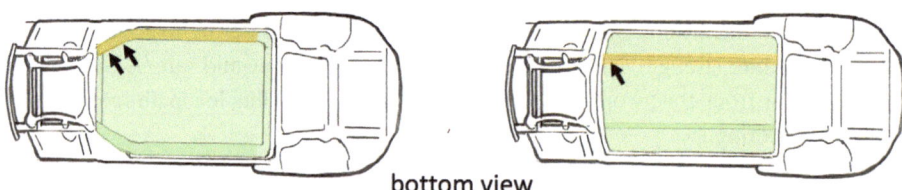

Fig. 4.15 Topology design elements, bottom view

Mass and Mass Distribution—Effect on Back-Up Structure Load Capacity

A BEV in the era of 600 Wh/l battery cells is heavier than its ICE counterpart and the vast majority of this difference lays in the propulsion battery. A BEV's propulsion battery can range from 400 kg (880 lbs) to over 1300 kg (2870 lbs). *To put the latter in perspective, consider that a 2023 Honda Civic also weighs 1300 kg.* As the propulsion battery does not ground out on the barrier during front crash events, the energy and deceleration forces associated with its mass must be managed by the back-up structure. Additionally, many BEVs package their front drive motor such that it does not ground out on the barrier, where as an ICE front motor often does.

Comparing a front wheel drive BEV to its ICE counterpart, the BEV back-up structure might be required to manage 450 to 1350 kg (50 kg subtracted to compensate for the lack of a fuel tank in BEVs) and the inertial forces associated with that (Fig. 4.16).

Mass Distribution—Effect on Front Compartment Topology

We found in Chap. 2 that the center of gravity for a BEV with an underfloor propulsion battery configuration is notably lower than in an ICE. Also recall the structural fundamental that the most efficient loadpath between an applied load and a boundary condition will be a straight line between the two. Consideration of these two concludes that an efficient BEV structure will have greater reliance on the cradle loadpath for the front crashworthiness strategy than its ICE counterpart, as illustrated in Fig. 4.17. Of course, how much of

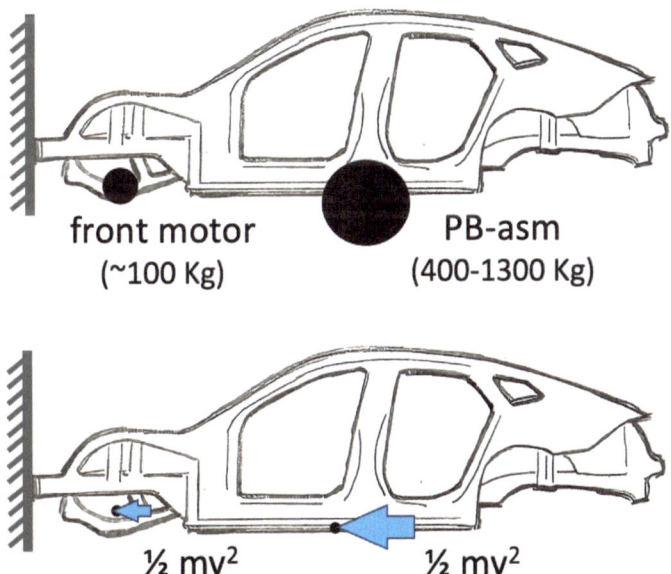

front motor   PB-asm
(~100 Kg)   (400-1300 Kg)

$\frac{1}{2}\,mv^2$ $\frac{1}{2}\,mv^2$

Fig. 4.16 EV ride-down components

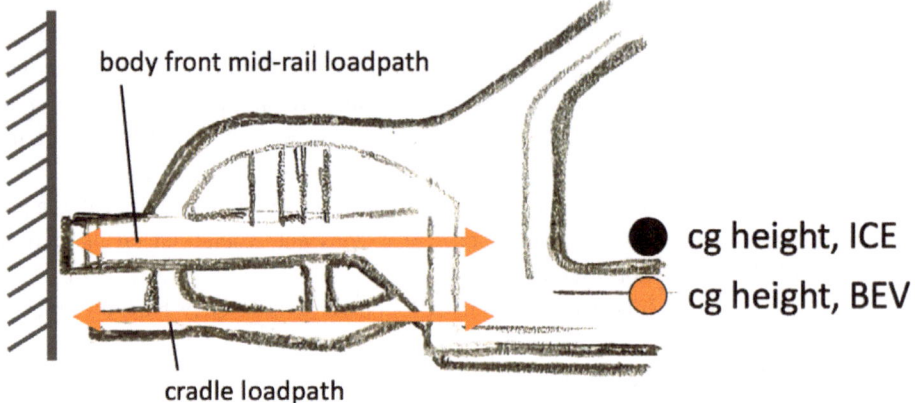

Fig. 4.17 Loadpath height

the crash load can be shifted to the cradle is dependent on many other things, including how strong the back-up structure at the cradle rear attachment can be made.

4.2 Moderate Overlap Frontal

The Moderate Overlap Frontal loadcase was introduced to represent car to car collisions where only a portion of the vehicle width is engaged; a condition where one vehicle drifts over the center line and collides with an oncoming vehicle. The overlap is less than the 100% overlap considered in the Full-Frontal loadcase and more than the 25% overlap of the Small Overlap Frontal, thus the descriptor "Moderate". The barrier is comprised by a deformable element mounted to a rigid foundation. The MOF test condition is illustrated below in Fig. 4.18.

This loadcase is found as a consumer metric and considered in several NCAPs around the globe. It is also an EU regulatory loadcase under the protocol UN R94.

The MOF loadcase is sometimes referred to as "front ODB", as the Barrier is Offset and Deformable.

The barrier overlap is defined as 40% of the vehicle's overall width (excluding side mirrors) as illustrated in Fig. 4.19.

Vehicle performance rating is based largely on occupant injury measurements taken from crash dummies but can include assessments of the structural performance. Figure 4.20 illustrates vehicle structure locations whose intrusion magnitude is considered in the IIHS performance rating protocol.

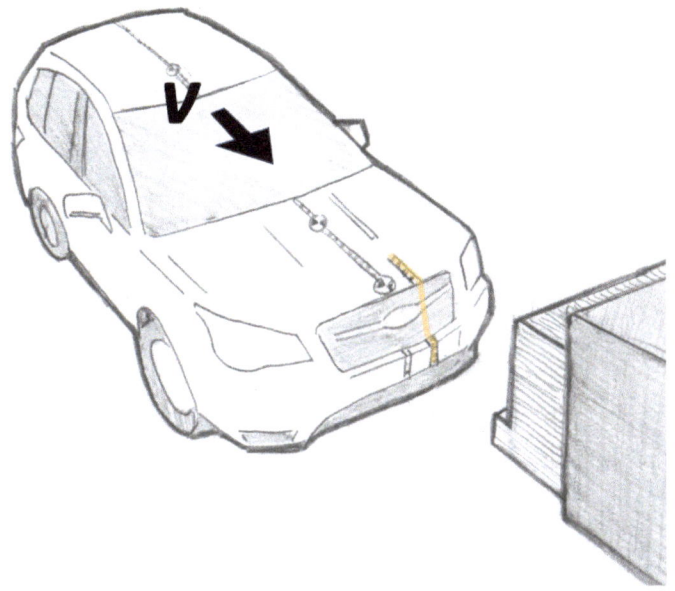

Fig. 4.18 General MOF test condition

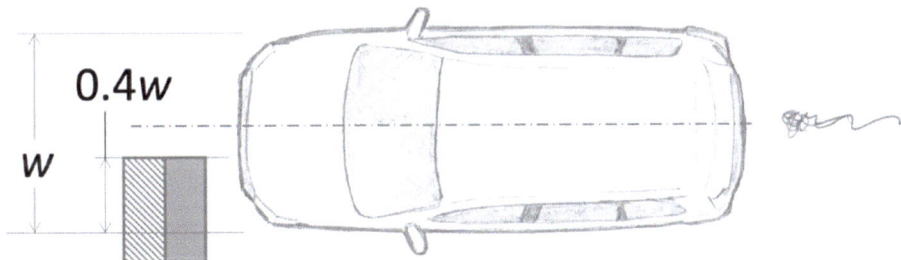

Fig. 4.19 MOF barrier overlap definition

Figure 4.21 illustrates that the MOF barrier overlap results in stable engagement of the near-side body *front-compartment-mid-rail*.

The video found through Link-4.5 provides an overview of the MOF loadcase.

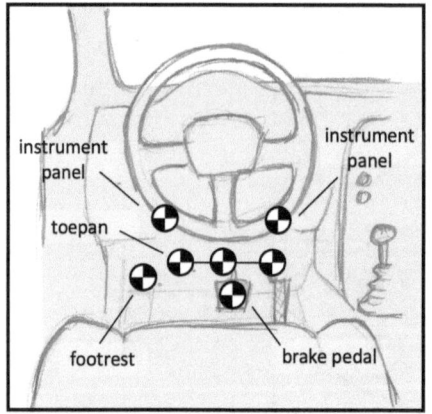 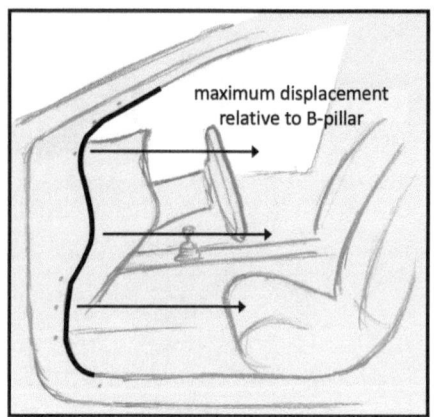

Fig. 4.20 IIHS-MOF structural intrusion measurement points

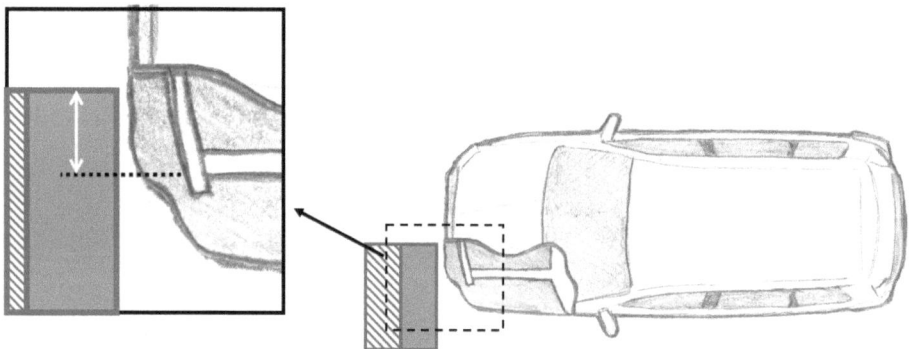

Fig. 4.21 Alignment of structure to MOF barrier

| 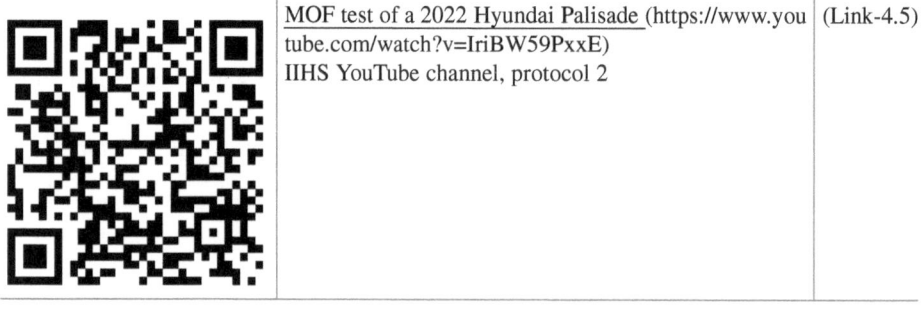 | MOF test of a 2022 Hyundai Palisade (https://www.you tube.com/watch?v=IriBW59PxxE) IIHS YouTube channel, protocol 2 | (Link-4.5) |

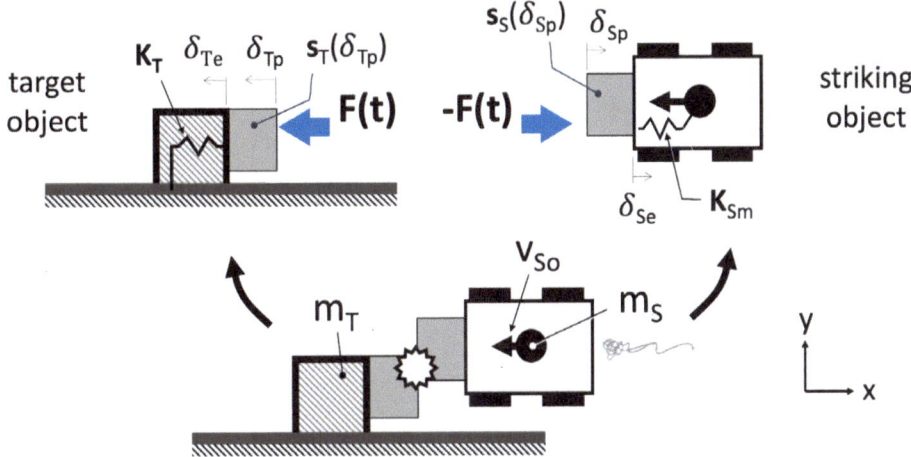

Fig. 4.22 General MOF collision model

4.2.1   Loadcase Modeling

The test vehicle approaches and strikes a stationary deformable barrier in this loadcase. Again, we can utilize the collision model developed in Sect. 3.2.3.5 by considering that the target object is essentially the barrier and the earth that it is attached to; $m_T \gg m_S$. The collision model is illustrated here in Fig. 4.22.

Fig. 4.22 variable definitions:

s_S	=	is the strength profile of the striking vehicle,
δ_{Sp}	=	is the direction of striking vehicle plastic deformation measure,
δ_{Se}	=	is the direction of striking vehicle elastic deflection measure
K_{Sm}	=	is the stiffness of the striking vehicle between impulse and mass center
s_T	=	is the strength profile of the barrier
δ_{Tp}	=	is the direction of barrier plastic deformation measure
δ_{Te}	=	is the direction of barrier elastic deflection measure
K_T	=	is the stiffness of the barrier

- Plasticity (s_T, s_S): Plasticity exists in both the striking and target objects.
- Friction: There will be a tendency for the striking object to rotate in a yaw direction during the event. The rotation would normally prompt us to include friction between the tires and ground in the model, however you will notice in the example video that the rotation largely occurs in the rebound stage of the event and while the wheels

are off the ground. It might be appropriate to consider tire friction if they leave skid marks during the collision stage of the event, however this is not typical for the MOF loadcase.

• Elasticity (K_{Sm}, K_T): Elasticity exists in both the barrier and striking object. The relevant stiffness of the striking vehicle is between the applied impulse force and its generalized center of mass.

Insight from Governing Equations

The impulse equations leveraged for our MOF collision model is the same as for the Full-Frontal loadcase and they are duplicated here. Impulse is defined as the change in a collision object's momentum (Eq. 4.26), but also equivalent to the area under the F(t) curve (4.27).

$$\text{Impulse}: \boxed{J = p_o - p_f} \qquad [\text{N s}]\,\text{or}\,[\text{kg m/s}] \qquad (4.26)$$

$$\text{Impulse}: \boxed{J = \int_o^f F(t)dt} \qquad (4.27)$$

or

$$\boxed{J = F_{ave}\Delta t} \qquad (4.28)$$

where:

J	=	impulse
p	=	momentum of a collision object
F	=	impulse force
Δt	=	time duration of collision

...equating Eqs. 4.28 and 4.26...

$$J = J \qquad (4.29)$$

$$F_{ave}\Delta t = p_o - p_f \qquad (4.30)$$

$$F_{ave}\Delta t = m_S v_{So} - m_S v_{Sf} \qquad (4.31)$$

$$F_{ave} \Delta t = m_S v_{So} - m_S(0) \tag{4.32}$$

$$\boxed{F_{ave} \Delta t = m_S v_{So}} \tag{4.33}$$

where:

m_S =   mass of the striking vehicle

v_{So} =   initial velocity of the striking vehicle

F_{ave} =   average impulse force

Δt =   time duration of collision.

As in the Full-Frontal loadcase, we see from Eq. 4.33 that the F·t product required to slow the striking vehicle is dependent on its mass; a heavier test vehicle will have more initial momentum and slowing it to a stop will require for 'force'.

Looking at the collision from the conservation of energy law…

$$E_f = E_o \tag{4.34}$$

$$\left(KE_f + W \right) = KE_o \tag{4.35}$$

…assuming that the striking vehicle's rotational KE at the end of test is insignificant and recognizing that assumption by using "≈"…

$$(0 + W) \approx \left(\frac{1}{2} m_S v_{So}^2 \right) \tag{4.36}$$

$$W = \frac{1}{2} m_S v_{So}^2 \tag{4.37}$$

$$\boxed{F_{ave} \left(\Delta_{Sp} + \Delta_{Tp} \right) \approx \frac{1}{2} m_S v_{So}^2} \tag{4.38}$$

where:

m_S =   mass of the striking vehicle

v_{So} =   initial velocity of the striking vehicle

v_{Sf} =   final velocity of the striking vehicle

F_{ave} =   average impulse force

Δ_{Sp} =   crush distance seen within the striking vehicle

Δ_{Tp} =   crush distance seen within the barrier

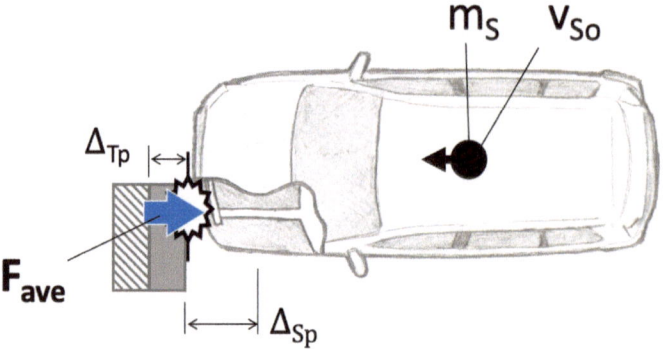

Fig. 4.23 MOF Conservation of energy FBD

Equation 4.38 is the similar to that found for the Full-Frontal loadcase (Eq. 4.13); the exception being that the crush distance in MOF is distributed between the striking object and the barrier. The striking vehicle's kinetic energy will be converted into material fracture and crush within the test vehicle and the barrier, as illustrated in Fig. 4.23. The amount of crush is dependent on the force capacity of the barrier and test vehicle; as the force capacities increase, the average impulse force increases and the amount of crush decreases.

Again, note that this equation is considered an approximate relationship because it does not consider the post-collision rotational velocity of the striking vehicle and the associated energy.

System Energy Flow

The initial system energy in MOF is converted into barrier crush ($F_{ave}\Delta_{Tp}$), crush in the striking vehicle ($F_{ave}\Delta_{Sp}$), and residual kinetic energy of the striking vehicle in the form of rotational velocity. Some energy is stored within the barrier and striking vehicle during the event. The amount of energy is dependent on the barrier and striking vehicle stiffnesses (K_T and K_{Sm}, respectively) and the maximum elastic compression occurring within these two objects (Δ_{Te} and Δ_{Se}, respectively). This energy is released once the impulse force is removed and 'springs' the striking vehicle off the barrier.

Figure 4.24 illustrates the system energy flow.

4.2.2   Fundamental Behavior and Physics

Figure 4.25 illustrates the FBD of the striking vehicle. The collision force will be resisted only by the striking vehicle mass, whose inertia will be changed during the event.

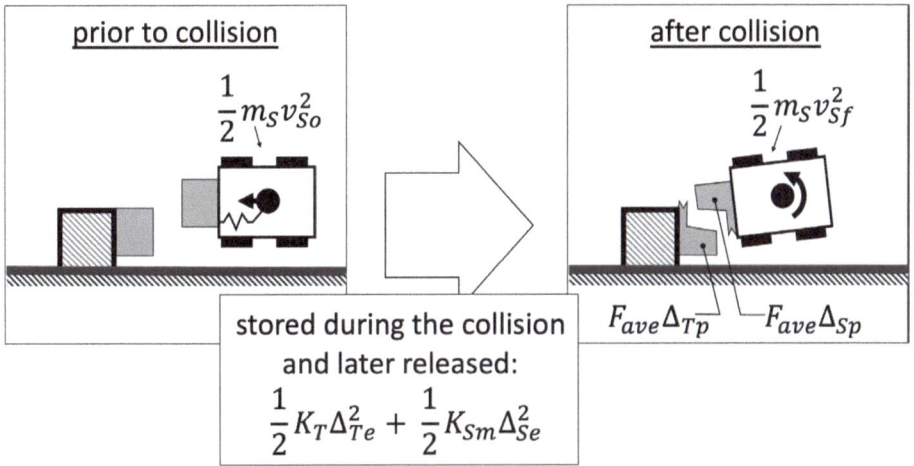

Fig. 4.24 MOF energy conservation

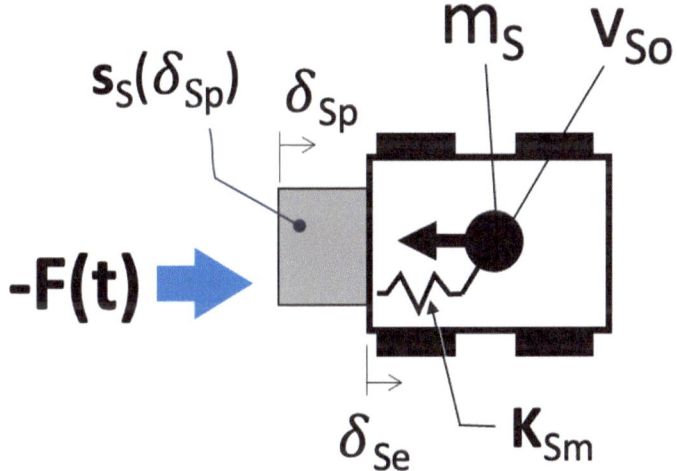

Fig. 4.25 MOF test vehicle FBD

Behavior of the striking vehicle in MOF is similar to that in other frontal loadcases, however the overlap condition and deformable characteristic of the barrier will cause differences.

Event Milestones

The behavior of the striking vehicle may vary depending on its weight and the parameters of the specific test protocol (particularly the barrier strength profile, barrier depth, and the striking vehicle speed), however the general behavior is illustrated in Fig. 4.26 and described below.

Behavior, High-Level
Stage-1: Crush Within the Striking Vehicle and the Barrier

(a) The striking vehicle first makes contact with the barrier face.
(b) Crush occurs within the barrier face and the striking vehicle. Crush in the striking vehicle occurs in the weakest portions of the load capability profile. Elastic compression of the striking vehicle begins as the impulse force is applied.

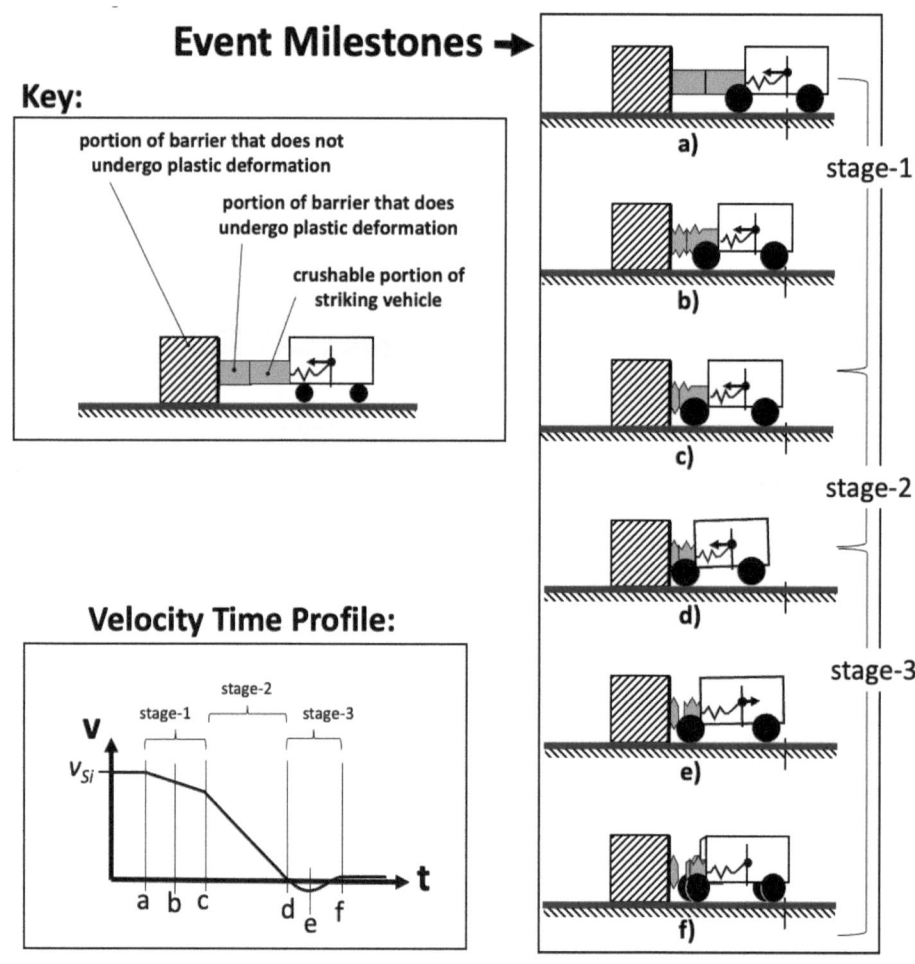

Fig. 4.26 MOF event milestones

Stage-2: Full Crush Achieved in the Barrier; Further Crush Occurs Only Within the Striking Vehicle

(c) There will be cases in which the striking vehicle still has velocity and kinetic energy after the barrier has been fully crushed. Again, depending on the striking vehicle mass and strength profile and the striking vehicle speed and barrier strength profile identified in the test protocol. In these cases, the striking vehicle engages the 'rigid' portion of the barrier and the impulse force increases. Crush now occurs in higher strength portions of the striking vehicle's load capability profile. Often the loadpath from the barrier to the *wheel/tire* to the *front-body-hinge-pillar* and *rocker* is a significant loadpath. Energy storage within the striking vehicle increases as elastic compression continues.

Stage-3: Rebound

(d) The impulse event is over; full crush has been achieved and the striking vehicle no longer has forward velocity. Release of the elastic energy stored in the striking vehicle and barrier begins, propelling the striking vehicle off the barrier. There will be a pitch motion of the striking vehicle in cases where there is a height offset between the release of stored energy and its cg. There will also be a yaw rotation, as there is a lateral offset between the vehicle cg and the vehicle's contact with the barrier.

(e) The striking vehicle comes to rest with some level of plan-view rotation relative to its initial direction.

Behavior, Detailed

Stage-1: Crush within the striking vehicle and the barrier

Material fracture and crush within the striking vehicle occurs in many of the vehicle components that lay within the path of the barrier.

Zones related to crush and back-up structure behavior can be defined for MOF as defined in the Full-Frontal loadcase, illustrated here in Fig. 4.27. These definitions are typically the same as in Full-Frontal, as deformation in the occupant compartment is not typically desired.

Also recall the concept of a structure's load capacity profile, from Sect. 4.1. Each loadpath within the structure will have a load capacity profile based on the strength of its individual segments. The body *front-compartment-mid-rail*, for example, tends to have an intentionally weaker segment at the front (S_{CZ1}), a slightly stronger crush zone rear of that (S_{CZ2}), and a stronger segment rear of that. Each individual loadpath is "backed up" by a strong, unyielding portion of the vehicle structure (S_{BUS}). Figure 4.28 illustrates the load capacities within the front structure loadpath segments.

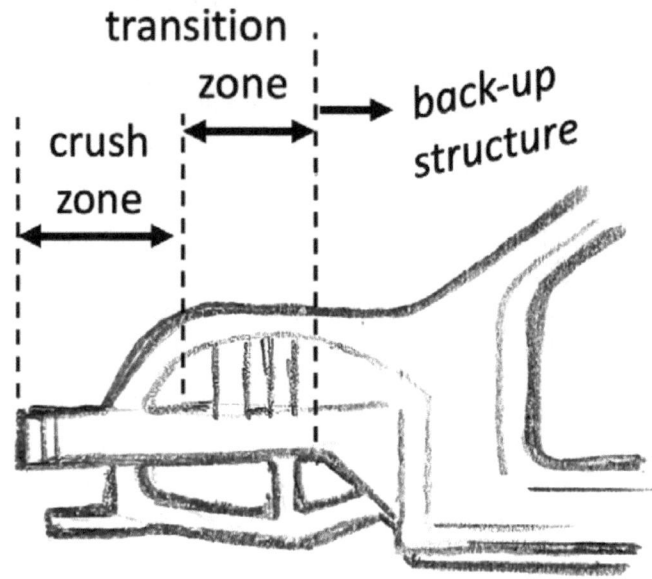

Fig. 4.27 Zones of the front structure

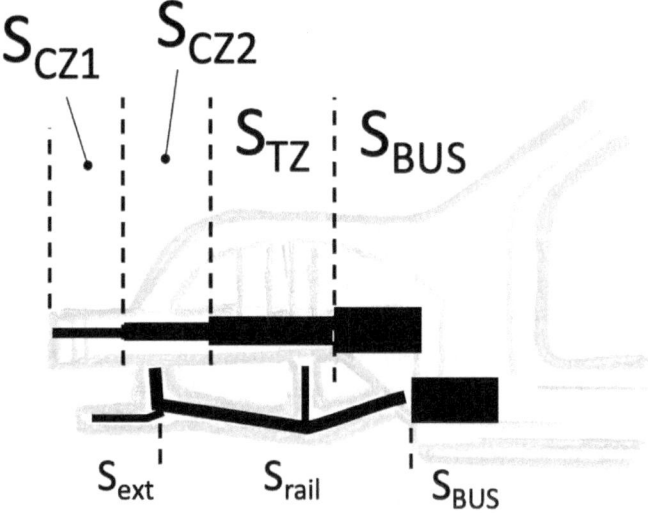

Fig. 4.28 Strength profile of the front structure

Stage-1 crush will begin at the weaker portions of the structure and barrier and progress to stronger portions.

Stage-2: Full Crush Achieved in the Barrier; Further Crush Occurs Only Within the Striking Vehicle

The barrier becomes much stiffer once the deformable portion is fully crushed and thus the impulse force magnitude will increase. Crush will now occur in stronger portions of the test vehicle.

'Shorting components' can modify structural loadpaths as the event progresses and we see an example of that in the Hyundai Palisade MOF video below (Link 4.6). Here, the front near-side wheel contacts the *front-body-hinge-pillar* during stage-2 while the striking vehicle still has velocity and kinetic energy. The strong wheel is captured between the barrier and the *front-body-hinge-pillar* and disables any crush that might otherwise occur in the corresponding segment of the vehicle structure. The wheel transmits some of the impulse force from the barrier to the *front-body-hinge-pillar*, as illustrated in Fig. 4.29. We can therefore expect some level of crush in this area and identify a portion of the *front-body-hinge-pillar* and the *rocker* as back-up structure.

 MOF test of a 2022 Hyundai Palisade (https://www.you tube.com/watch?v=IriBW59PxxE) IIHS YouTube channel, protocol 2 (Link-4.6)

As in the Full-Frontal loadcase, crush is important for vehicle performance. The following relationships between crush and the structural implications or performance derived in the Full-Frontal section are also applicable here.

(1) An approximate relationship between crush and rail force derived using the conservation of energy law, Eq. 4.15…

$$F_{rail} \approx \frac{\frac{1}{2} m_S v_{So}^2}{\Delta_{Sp}}$$

(4.39)

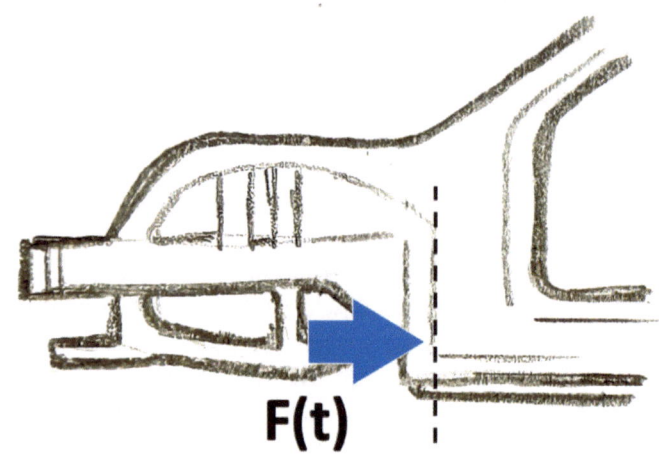

Fig. 4.29 Front-body-hinge-pillar loadpath

where:

F_{rail} = force seen by the body mid-rail

v_{So} = initial velocity of the striking vehicle

F_{ave} = average impulse force

Δ_{Sp} = crush distance seen within the striking vehicle

   Note that this is an approximate relationship, as it considers a loadpath consisting only of the mid-rail; no cradle participation is considered, for example. The equation's accuracy reduces further since components short the crush during the event and introduce additional loadpaths.

(2) The relationship between the generalized force applied to the rail and the time duration of crush derived from Newton's second law of motion equation; Eq. 4.19…

$$F_{rail} \approx m_s \frac{v_{so}}{\Delta t} \qquad (4.40)$$

where:

F_{rail} = force seen by the body mid-rail

m_S = mass of the striking vehicle

a_S = deceleration of the striking vehicle

v_{So} =   initial velocity of the striking vehicle

v_{Sf} =   final velocity of the striking vehicle

Δt =   time duration of the impulse event

Like in the Full-Frontal loadcase, the characteristics of structural deceleration has implications on occupant performance. Section 3.3 details the 'Other Collision' and can be referenced for more information.

The above equations can have some limited practical purpose in early design development but they are provided here predominantly to help develop intuition.

Stage-3: Rebound

The energy stored within the barrier and striking vehicle is released as the impulse force is reduced and withdrawn; the mechanism of which is detailed in Sect. 3.2.3.2. This energy release causes the striking vehicle to 'spring off' of the barrier at the end of the collision event.

A test vehicle in the MOF loadcase will exhibit a yaw motion during the collision event due to the misalignment between the impulse force and the vehicle's cg, as illustrated in Fig. 4.30.

4.2.3   Structural Loadpath Topology

As the same components engage the barrier, it's no surprise that the loadpath topology for MOF has similarities to that of the Full-Frontal loadcase. Structural components of the front compartment whose role is energy absorption are included in the topology; the *front-impact-beam*, body *front-compartment-mid-rails*, and *front-cradle*. Since only half of the front structure manages the collision energy and the front near-side wheel often impacts the *front-body-hinge-pillar* in MOF, the forces internal to the structure will be higher than in the Full-Frontal loadcase. We find that the energy ride-down portion of the MOF loadpath topology includes the *front-body-hinge-pillar*, *center-compartment-mid-rail*, *rocker*, and *A-pillar*.

The behavior of the 1995 Chrysler Cirrus in the IIHS's initial MOF test illustrates this. The Cirrus's design predates the introduction of the MOF loadcase and consideration of this loadcase in its design development process is unlikely. As the video referenced by Link 4.7 shows, the Cirrus exhibited buckling in the *A-pillar* and more crush in the lower *front-body-hinge-pillar* than a Good MOF rating would allow.

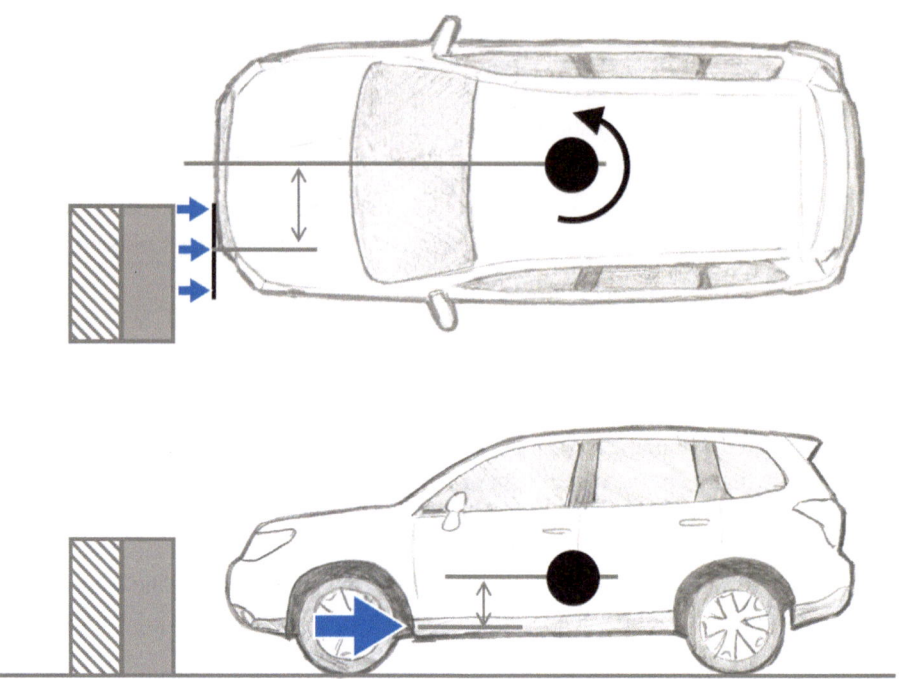

Fig. 4.30 Vehicle cg misalignment to Barrier and Yaw Motion

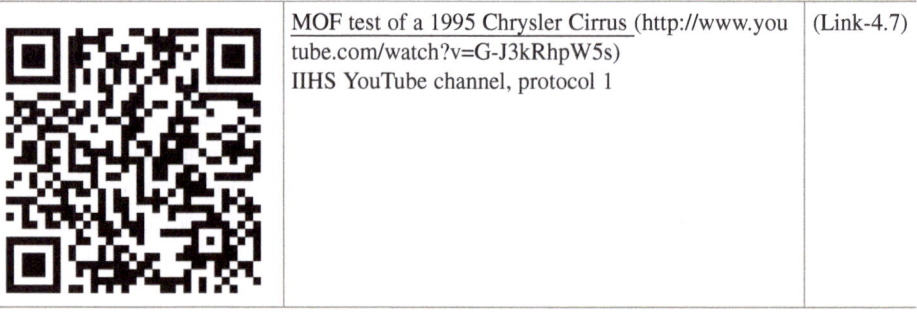

	MOF test of a 1995 Chrysler Cirrus (http://www.you tube.com/watch?v=G-J3kRhpW5s) IIHS YouTube channel, protocol 1	(Link-4.7)

As might be expected, reinforcement to the bodyside structure was introduced as manufacturers considered the MOF loadcase in their vehicle designs.

Because of MOF engages only one *front-compartment-mid-rail* and both are engaged in the Full-Frontal loadcase, it can be challenging to engineer the *mid-rail* force capacity and crush behavior such that it performs well in both conditions.

The general structural loadpath topology for MOF is illustrated in Fig. 4.31. Note that it can be difficult to define a threshold for what structure is highlighted, as structural

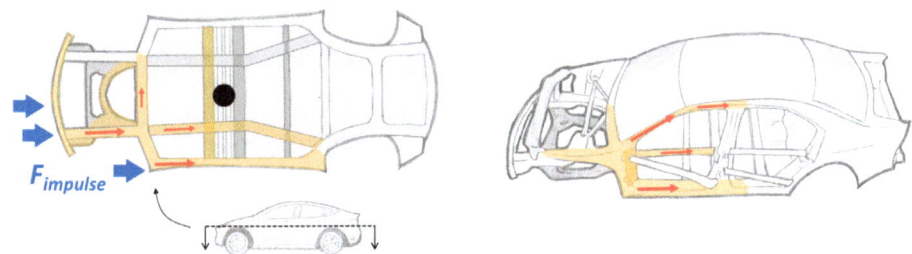

Fig. 4.31 Typical MOF structural loadpath topology

participation of the segments is not a digital attribute. You will find that the bounds will be different depending on vehicle attributes and behavior.

4.2.4 Peculiarities of EVs

Post-Crash High Voltage Safety

Recall the overview of FMVSS305 and UNIECE regulations provided in this chapter's introduction. These regulations apply to vehicles with high voltage content and specify post-crash performance related to; protection from electrical shock, level of electrolyte spillage, and retention of the propulsion battery housing. The UNECE regulation applies to the EN R94 MOF loadcase.

Like described in the Full-Frontal section, packaging of high-voltage components requires thoughtful consideration to efficiently and robustly satisfy the demands of the forementioned regulations. It is preferable to package high voltage content in areas of the structure that are free from crush and large elastic displacements.

Mass and Mass Distribution

A BEV in the era of 600 Wh/l battery cells is heavier than its ICE counterpart, the vast majority of this difference is due to the presence of the propulsion battery. As detailed

Fig. 4.32 General MPDB test condition

in the Full-Frontal section, this condition requires additional back-up structure strength, relative to the BEV's ICE counterpart.

4.3   Front Moving Progressive Deformable Barrier

The configuration of the Front Moving Progressive Deformable Barrier (MPDB) loadcase represents car to car collisions where only a portion of the vehicle width is engaged; a condition where one vehicle drifts over the center line and collides with an oncoming vehicle. MPDB is similar to the MOF loadcase, however its setup and protocol are very different. The MPDB barrier is deformable, as in MOF, however it is mounted to a moving trolly in MPDB (thus the descriptor "moving"). At the time of this book's writing, all protocols specify that the test vehicle and the MDB have the same initial speed (50 kph / 31 mph) however they are traveling in opposite directions. The strength of the MPDB barrier increases as the crush magnitude increases, thus the descriptor "progressive". The MPDB test condition is illustrated below in Fig. 4.32.

MPDB was introduced primarily to assess how 'friendly' or 'unfriendly' a vehicle's crush behavior is to the other vehicle involved in the collision and to encourage manufacturers to design automotives that are 'compatible' with others. 'Friendliness' is determined by measuring how uniform the barrier deformation is. Barrier deformation that is more uniform and distributed within the barrier face surface indicates that the test vehicle will make it easier for the opposing vehicle to achieve good performance while barrier deformation that is highly localized indicates that the test vehicle will promote poor performance in the other vehicle.

In addition to an assessment of barrier deformation uniformity, assessment of occupant injury metrics and occupant compartment intrusion measures are included in the vehicle performance rating. This loadcase is a part of several NCAPs around the globe, including in Europe, Japan, and China.

The barrier overlap is defined as 50% of the vehicle's overall width (excluding side mirrors) as illustrated in Fig. 4.33 and seen by the video referenced by Link 4.8.

|  | MPDB test of a 2021 Nio ES8 (https://www.youtube.com/watch?v=c10z2q_GOs0) EuroNCAP YouTube Channel | (Link-4.8) |

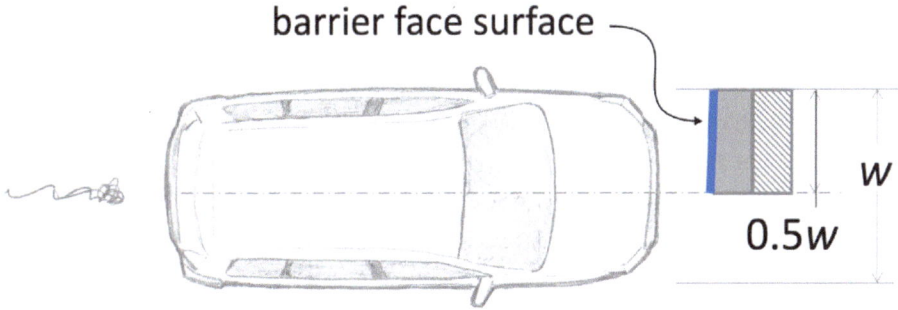

Fig. 4.33 MPDB barrier overlap definition

4.3.1 Loadcase Modeling

In this loadcase, the test vehicle approaches and strikes a deformable barrier which is moving towards it. Again, we can utilize the collision model developed in Sect. 3.2.3.5. An illustration of a MPDB collision model is shown in Fig. 4.34. We will consider the trolly to be the striking object and refer to the target object as the 'test vehicle'.

Fig. 4.34 variable definitions:

s_T = is the strength profile of the test vehicle

δ_{Tp} = is the direction of test vehicle plastic deformation measure

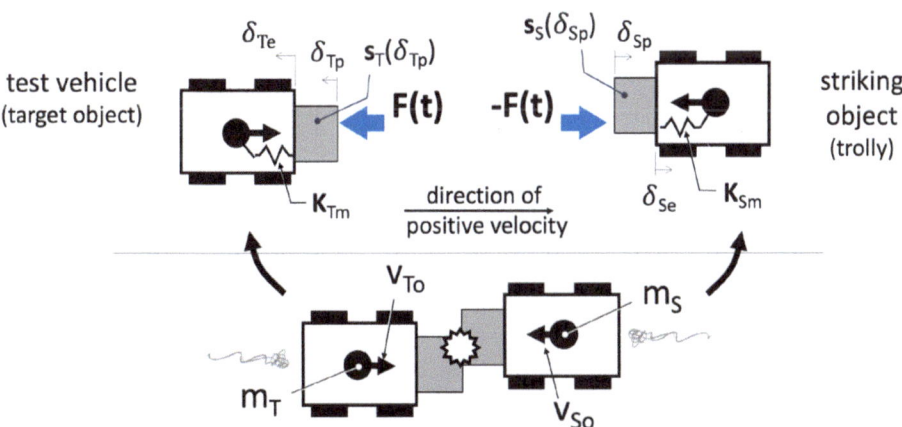

Fig. 4.34 General MPDB collision

δ_{Te} =   is the direction of test vehicle elastic deflection measure

K_{Tm} =   is the stiffness of the test vehicle between impulse and mass center

s_S =   is the strength profile of the striking vehicle

δ_{Sp} =   is the direction of striking vehicle plastic deformation measure

δ_{Se} =   is the direction of striking vehicle elastic deflection measure

K_{Sm} =   is the stiffness of the striking vehicle between impulse and mass center

- Plasticity (s_T, s_S): Plasticity exists in both the test and striking vehicles.
- Friction: The example video link-4.8 shows how the two objects rotate in plan-view during the event. Rotation would normally prompt us to include friction between the tires and ground in the model, however rotation occurs after maximum crush has been achieved and we will therefore ignore friction.
- Elasticity (K_{Sm}, K_{Tm}): Elasticity exists in both the vehicle. In each vehicle, the relevant stiffness is between the impulse force and the generalized center of mass.

Insight from Governing Equations

It is useful to understand how the impulse magnitude responds to changes in the test vehicle mass, and we can determine that by starting with the impulse equations below; which define impulse as the change in a collision object's momentum (Eq. 4.41) and equivalent to the area under the F(t) curve (Eq. 4.42).

$$\text{Impulse}: \quad \boxed{J = p_o - p_f} \qquad \text{[N s] or [kg m/s]} \qquad (4.41)$$

$$\text{Impulse}: \quad \boxed{J = \int_o^f F(t)\,dt} \qquad\qquad (4.42)$$

or

$$\boxed{J = F_{ave}\Delta t} \qquad\qquad (4.43)$$

where:

J =   impulse

p =   momentum of one collision object

F =   impulse force

Δt =   time duration of collision

... equating these two with a focus on the test vehicle...

$$J = J \tag{4.44}$$

$$F_{ave}\Delta t = p_o - p_f \tag{4.45}$$

$$F_{ave}\Delta t = m_T v_{To} - m_T v_{Tf} \tag{4.46}$$

$$\boxed{F_{ave}\Delta t = m_T \left(v_{To} - v_{Tf} \right)} \tag{4.47}$$

where:

m_T	=	mass of the test vehicle
v_{To}	=	initial velocity of the test vehicle
v_{Tf}	=	velocity of the test vehicle
F_{ave}	=	average impulse force
Δt	=	time duration of collision

Equation 4.47 shows that the impulse magnitude $F_{ave}\Delta t$ increases as the mass of the test vehicle increases and the difference between its starting and final velocity increases. Further progression of the equation requires an understanding of how the velocity change is affected by the input variables.

Starting with the conservation of momentum equation ...

$$\text{total system} : \sum p_f = \sum p_o \tag{4.48}$$

$$m_T v_{Tf} + m_S v_{Sf} = m_T v_{To} + m_S v_{So} \tag{4.49}$$

...the vehicles have the same speed at the beginning of the event, however the striking vehicle is traveling in the negative direction of travel, so...

$$v_{To} = v_o \tag{4.50}$$

$$v_{So} = -v_o \tag{4.51}$$

...the vehicles will have the same speed at the end of the event, so...

$$v_{Tf} = v_{Sf} = v_f \tag{4.52}$$

...substituting (4.51) and (4.52) into (4.49) yields,

$$v_f(m_T + m_S) = m_T v_o - m_S v_o \tag{4.53}$$

$$v_f(m_T + m_S) = v_o(m_T - m_S) \tag{4.54}$$

...and solving for the final velocity...

$$\boxed{v_f = v_o\left(\frac{(m_T - m_S)}{(m_T + m_S)}\right)} \tag{4.55}$$

Here we see that the final velocity is the initial velocity multiplied by a mass ratio. Equation 4.55 also indicates that there are three distinct conditions depending on the mass of the two objects, as illustrated by Table 4.1. In cases where the test vehicle's mass is heavier than the striking vehicle, the test vehicle will end the test traveling in the same direction it started in. In the case where the test and striking vehicles have the same mass, their final velocity will be zero; both vehicles coming to a rest at the end of test. And finally, in cases where the test vehicle's mass is less than the striking vehicle's, the final velocity will be in the opposite direction of its initial condition.

- Note that these behaviors are consistent with those described in Table 3.3.
- As a reference, the striking vehicle mass is 1400 kg in the EuroNCAP test protocol.

Table 4.1 MPDB Vehicle behavior based on mass

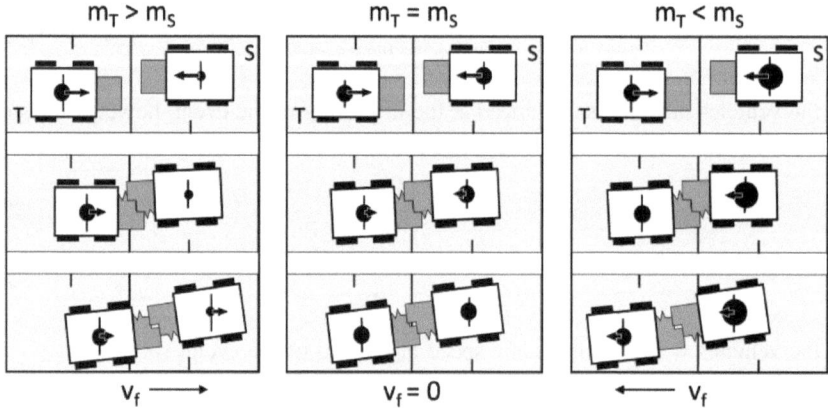

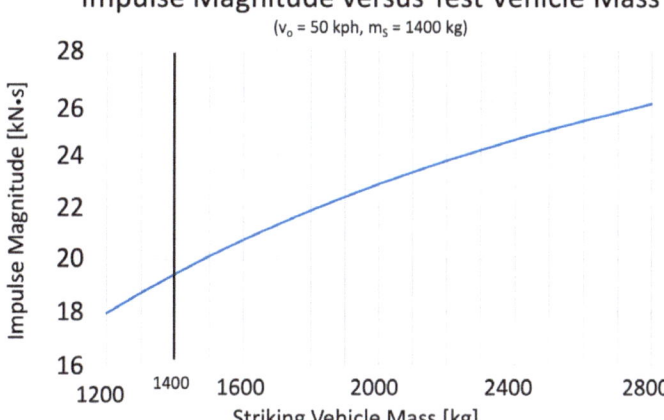

Fig. 4.35 Impulse magnitude versus test vehicle mass

Back to the question of how the impulse magnitude is affected by the test vehicle mass, we can substitute Eq. 4.55 into 4.47, which yields...

$$F_{ave}\Delta t = m_T \left(v_o - v_o \left(\frac{(m_T - m_S)}{(m_T + m_S)} \right) \right) \tag{4.56}$$

Equation 4.56 shows a relationship between the striking vehicle mass and the impulse magnitude, however it is easier to see in graphical form, as shown in Fig. 4.35. The impulse magnitude increases linearly as the mass of the striking vehicle increases.

Another question is how Work responds to the input variables; specifically, **how is the crush magnitude affected by changing input variables?** We know that the total system energy increases as the test vehicle mass increases, but how is that energy distributed between crush and residual kinetic energy?

Looking at the collision from the conservation of energy law...

$$E_f = E_o \tag{4.57}$$

$$\left(KE_f + W \right) = KE_o \tag{4.58}$$

$$W = KE_o - KE_f \tag{4.59}$$

$$W = \left(\frac{1}{2}m_T v_{To}^2 + \frac{1}{2}m_S v_{So}^2 \right) - \left(\frac{1}{2}m_T v_{Tf}^2 + \frac{1}{2}m_S v_{Sf}^2 \right) \tag{4.60}$$

...knowing that both vehicles have the same initial speed and final velocity (we can ignore the fact that the striking vehicle's initial velocity is negative because the velocity terms are squared in this equation)...

$$W = \left(\frac{1}{2}m_T v_o^2 + \frac{1}{2}m_S v_o^2\right) - \left(\frac{1}{2}m_T v_f^2 + \frac{1}{2}m_S v_f^2\right) \tag{4.61}$$

$$W = \frac{1}{2}\left(v_o^2(m_T + m_S) - v_f^2(m_T + m_S)\right) \tag{4.62}$$

...substituting Eq. 4.55 into Eq. 4.62...

$$W = \frac{1}{2}\left(v_o^2(m_T + m_S) - \left(v_o\frac{(m_T - m_S)}{(m_T + m_S)}\right)^2(m_T + m_S)\right) \tag{4.63}$$

...manipulation simplifies the equation to...

$$W = \frac{1}{2}\left(v_o^2\left((m_T + m_S) - \frac{(m_T - m_S)^2}{(m_T + m_S)}\right)\right) \tag{4.64}$$

...and finally, substituting the definition of work (Eq. 4.65) into Eq. 4.64...

$$W = F_{ave}(\Delta_{Sp} + \Delta_{Tp}) \tag{4.65}$$

$$F_{ave}(\Delta_{Sp} + \Delta_{Tp}) = \frac{1}{2}\left(v_o^2\left((m_T + m_S) - \frac{(m_T - m_S)^2}{(m_T + m_S)}\right)\right) \tag{4.66}$$

where:

m_T	=	mass of the test vehicle
m_S	=	mass of the striking vehicle
v_{To}	=	initial velocity of the test vehicle
v_{So}	=	initial velocity of the striking vehicle
v_f	=	final velocity of the vehicles
F_{ave}	=	average impulse force
Δ_{Tp}	=	crush distance seen within the test vehicle
Δ_{Sp}	=	crush distance seen within striking vehicle

Equation 4.66 is somewhat of a monster. We can see, however, that the work done by crush will be related to the initial velocity and a vehicle mass ratio. Figure 4.36 shows the relationship in graphical form, showing that work increases as the test vehicle mass increases and that work is on the order of 280–340 kJ.

Distribution of crush between the MDB and the striking vehicle will depend in their relative strength profiles, $s_T(\delta_{T_p})$ and $s_S(\delta_{S_p})$.

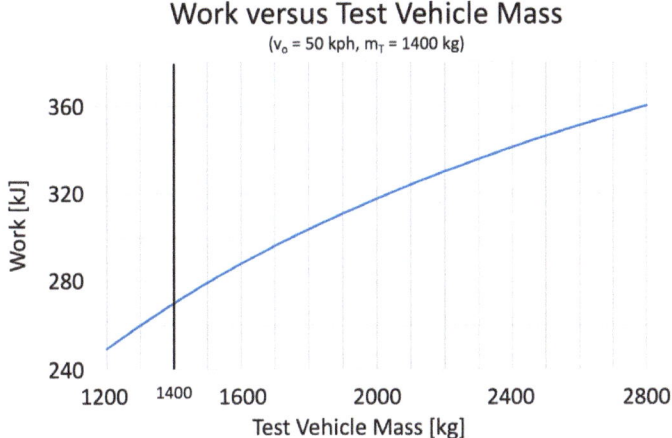

Fig. 4.36 Work versus test vehicle mass

System Energy Flow
The initial system energy in MPDB is distributed between barrier crush ($F_{ave}\Delta T_p$), crush in the test vehicle ($F_{ave}\Delta S_p$), and residual kinetic energy of the two vehicles in the form of translational and rotational velocity. Some energy is stored during the event, the amount of which is dependent on the stiffness properties of the two vehicles (K_{Sm}, K_{Tm}) and the maximum elastic compression seen within each (ΔS_e, ΔT_e). This energy is released once the impulse force is removed and results in the two vehicles 'springing' apart. Figure 4.37 illustrates the system energy flow. Note that the final velocity of the test vehicle is shown in the same direction of its initial travel, representing the condition where $m_T > m_S$.

4.3.2   Fundamental Behavior

Figure 4.38 illustrates the FBD of the test vehicle. The collision force will be resisted only by its mass, whose inertia will be changed during the event.

Behavior of the test vehicle in MPDB is similar to that in the MOF loadcase, however differences in the overlap, barrier characteristics, and vehicle speed will cause differences.

Event Milestones
The general behavior of the vehicles in MPDB is illustrated in Fig. 4.39. The velocity time profile and time step images represent the case where $m_T > m_S$.

Behavior, High-Level

(a) The two vehicles first make contact.

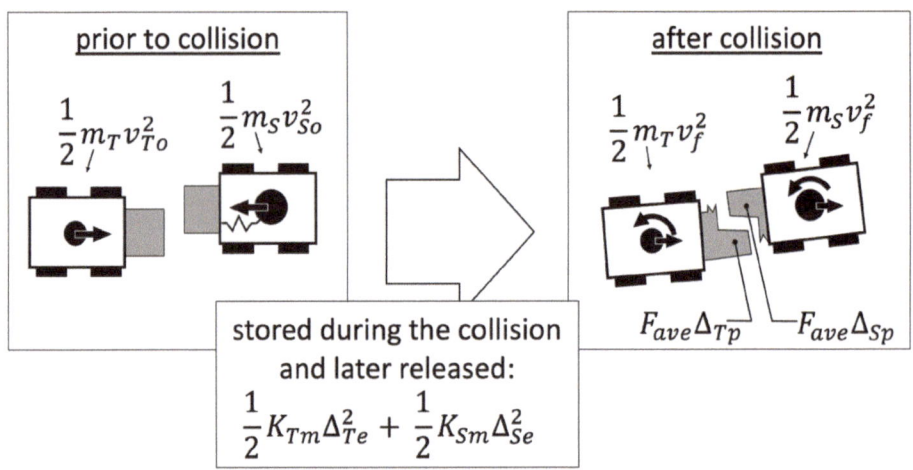

Fig. 4.37 MPDB energy conservation

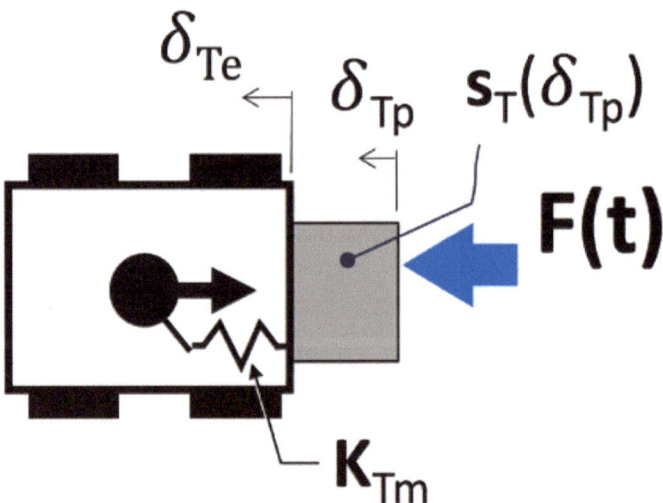

Fig. 4.38 MPDB test vehicle FBD

(b) Crush occurs within the barrier face and the test vehicle. Crush in the test vehicle occurs in the weakest portions of the load capability profile. Elastic compression in both vehicles begins as the impulse force is applied.

(c) The instantaneous velocity of the lighter vehicle is zero and crush continues if $m_T \neq m_S$.

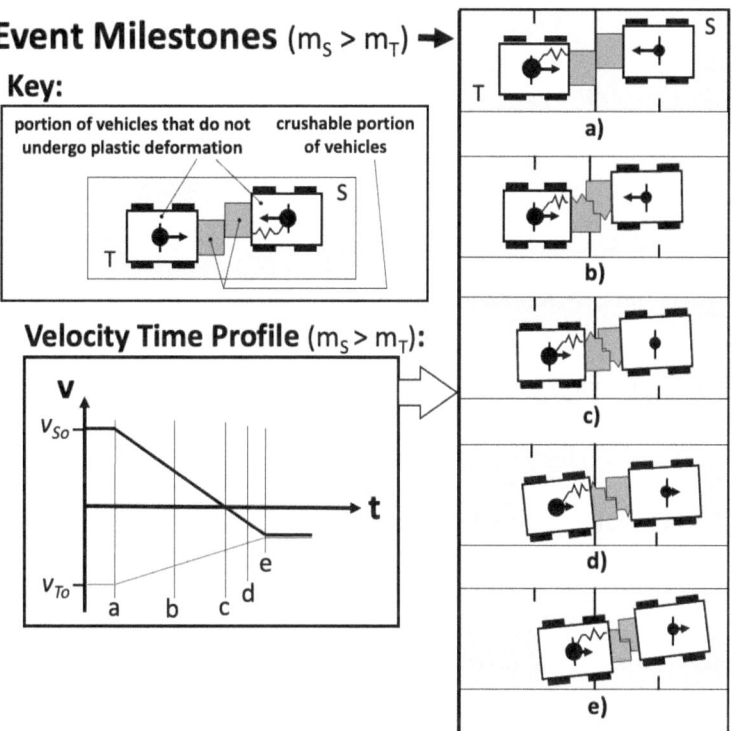

Fig. 4.39 MPDB event milestones

(d) Crush continues and the lighter vehicle accelerates in the opposite direction of its initial velocity.

(e) The impulse event is over; full crush has been achieved. The two vehicles move together at the same speed if $m_T \neq m_S$. Stored elastic energy has been released, although visual indications of a spring-back will likely be difficult to observe due to the interlocking nature of crush. Both vehicles will exhibit yaw motion as there is a lateral offset between each vehicle's cg and the impulse force. Either or both vehicles will also exhibit a pitch motion when there is a height offset between the location of the impulse and their cg.

Behavior, Detailed

Detailed behavior of the test vehicle during the MPDB crush stage is extremely similar to that of the striking vehicle in the MOF loadcase. Refer to "Stage-1: Crush within the

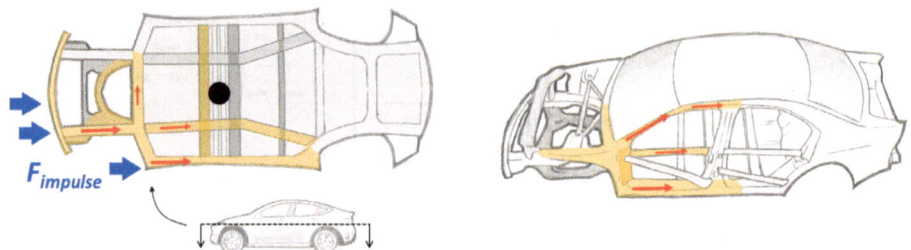

Fig. 4.40 Typical MPDB structural loadpath topology

striking vehicle and the barrier" for a deeper understanding of test vehicle behavior in MPDB.

4.3.3 Structural Loadpath Topology

As the same components engage the barrier, it's no surprise that the loadpath topology for MOF is essentially the same as for the MOF loadcase, illustrated in Fig. 4.40.

Assessment of the barrier deformation uniformity introduces another significant aspect of performance development; motivation to distribute the impulse across the surface of the barrier. Increasing the surface area of the structure that contacts the barrier and tuning the relative strength of the contacting structural elements are a typical design elements incorporated for MPDB performance. There is a higher motivation to increase the structural surface area and tune loadpath strength profiles as the test vehicle mass increases; as the impulse force increases as a function of test vehicle mass, (Fig. 4.35).

Figure 4.41 illustrates a *front-impact-beam* with a larger frontal area and the addition, or strengthening, of a *lower-impact-beam*.

4.3.4 Peculiarities of EVs

Post-Crash High Voltage Safety
Although the MPDB loadcase is not considered under the FMVSS305 and UNIECE regulations, automobile manufacturers exercise due-care and design for high voltage electrical integrity in other loadcases, MPDB included.

As described in the previous sections detailing frontal loadcases, ensuring post-crash high-voltage safety requires thoughtful packaging of high-voltage components with respect to crush zones and low displacement elastic deflections.

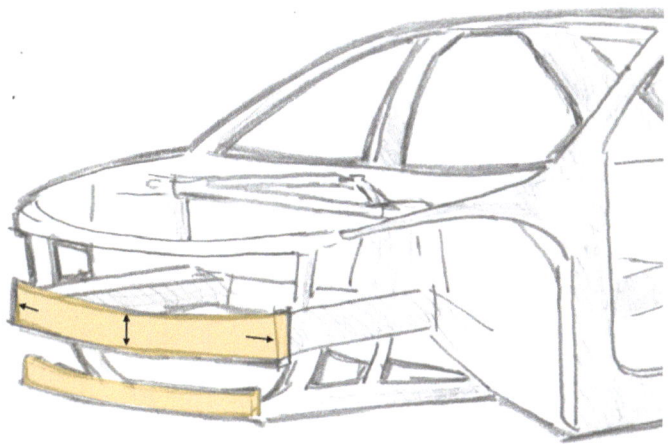

Fig. 4.41 Force distribution design elements

Mass and Mass Distribution

A BEV in the era of 600 Wh/l battery cells is heavier than its ICE counterpart and as described in the Full-Frontal section, this condition requires additional back-up structure strength relative to the BEV's ICE counterpart. Recall the relationship between impulse magnitude and striking vehicle mass shown in Fig. 4.35; the 600 kg difference between a 2024 Chevrolet Blazer ICE and a 2024 Blazer EV equates to a 30% increase in impulse magnitude.

4.3.5   Small Overlap Frontal

Small Overlap Frontal, or SOF, was introduced by the IIHS in 2012 after finding the condition significant in a study of front crash fatalities. In this test, the vehicle of interest is traveling at 64.4 kph (40 mph) and strikes a 'rigid' barrier, as illustrated by Fig. 4.42. Link 4.9 provides an example video.

The SOF barrier does not engage the entire width of the vehicle, as in the Full-Frontal loadcase, and has even less overlap than in the MOF or MPDB loadcases. The IIHS SOF protocol defines the barrier overlap as 25% of the vehicle's overall width (excluding side mirrors), as illustrated in Fig. 4.43. The IIHS defined barrier also has a 150 mm radius on its inside edge.

Basis for vehicle performance ratings include dummy measurements and displacements at specific vehicle locations. Vehicle locations measured in the IIHS protocol (version 7) are illustrated in Fig. 4.44.

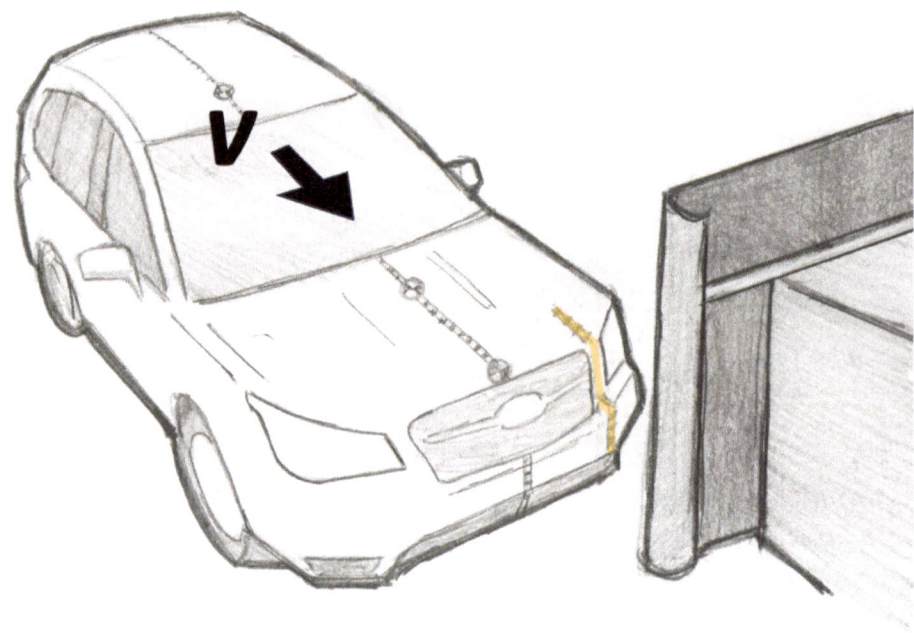

Fig. 4.42 General SOF test condition

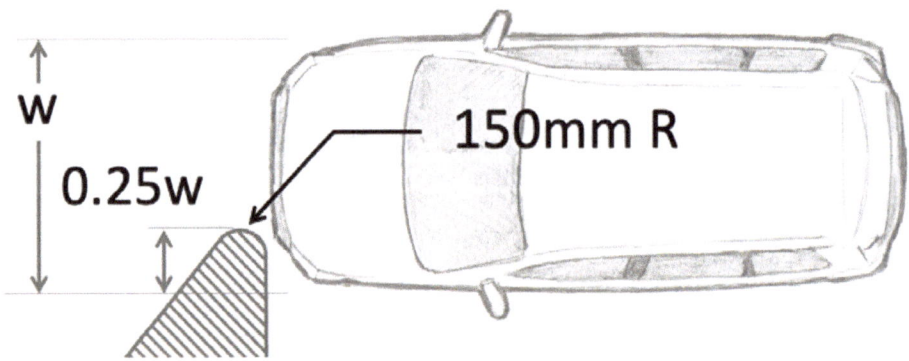

Fig. 4.43 SOF overlap definition

The SOF loadcase is sometimes referred to as "Small Overlap Rigid Barrier", or "SORB".

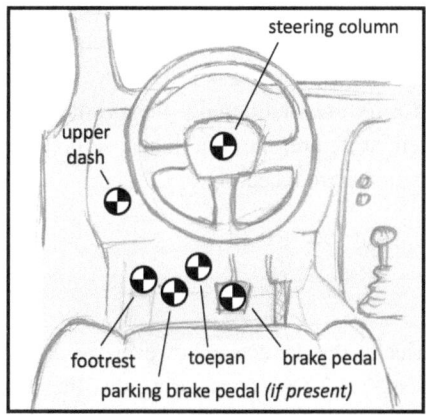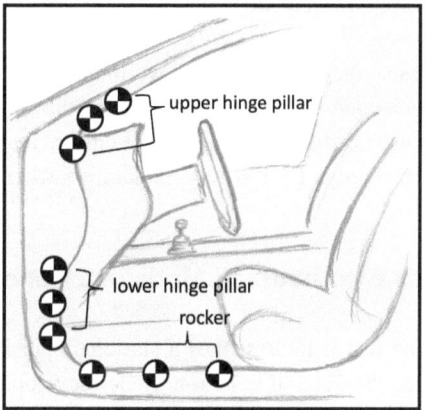

Fig. 4.44 IIHS-SOF structural intrusion measurement points

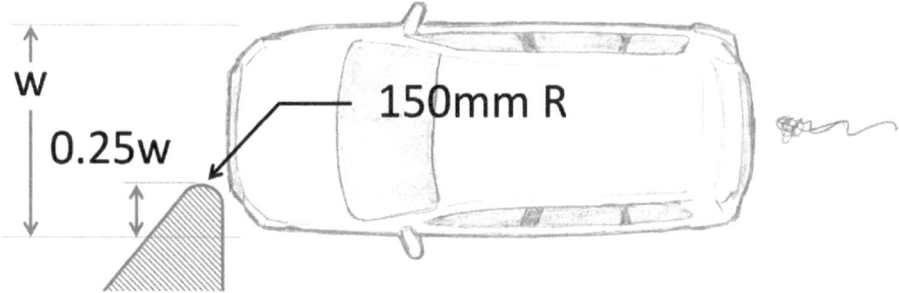

Fig. 4.45 Misalignment of structure to SOF barrier

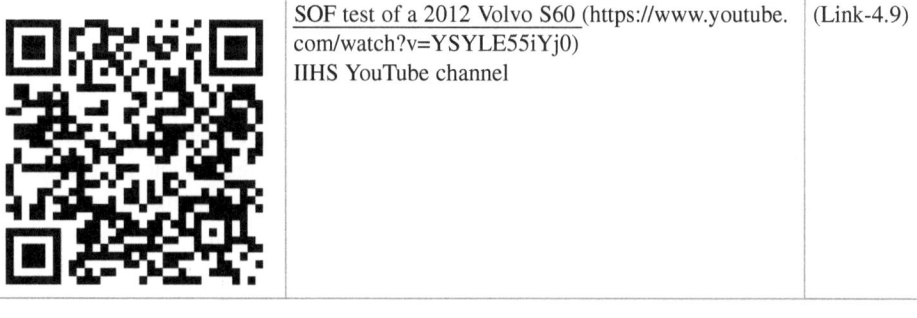

	SOF test of a 2012 Volvo S60 (https://www.youtube.com/watch?v=YSYLE55iYj0) IIHS YouTube channel	(Link-4.9)

Figure 4.45 illustrates that the overlap and barrier shape are such that the body *front-compartment-mid-rail* is typically not aligned to the flat portion of the barrier. In these cases, the *rail* does not have long-lasting barrier engagement and is usually not an effective longitudinal loadpath. Since the body *mid-rails* are traditionally the primary front crashworthiness loadpath and energy management structure, the SOF loadcase presented a challenge for most manufacturers when it was first introduced.

4.3.6 Loadcase Modeling, Generic

The test vehicle strikes a stationary 'rigid' barrier in this loadcase, however we can utilize the general collision model developed in Sect. 3.2.3.5 by considering that the target object in SOF is essentially the barrier and the earth that it is attached to; $m_T \gg m_s$. We would therefore expect unmeasurable velocity change of the barrier/earth and a significant amount of velocity change in the striking vehicle. Although the first is true, the second is not necessarily the case. Since the barrier does not have 100% overlap to the striking vehicle and there is a lateral offset between the collision surface and the striking vehicle's cg; some off-axis loading will occur and the striking vehicle will actually pass by the barrier in some cases. This turns out to have significant consequence.

The general collision model for SOF is shown here in Fig. 4.46.

- Plasticity (s_S): Plasticity exists only in the striking object, assuming that the barrier has been designed and constructed well. The scope and extent of vehicle crush depends largely on the amount of momentum change seen by the striking object.
- Friction (f): Friction at the tire that contacts the barrier is insignificant, as motion of this tire is arrested early in the event. Friction at the other tires comes into play in cases where the striking object exhibits significant lateral translation during the event. In these cases, the laterally translating tires leave skid marks which can be seen in crash test videos. Friction is definitely a part of the equation, however it can usually be ignored based on its relatively small influence on vehicle behavior and performance results.
- Elasticity (K_T, K_{Sm}, K_{Sf}): Although referred to as a "rigid", the barrier does have elasticity and is represented by K_T. Displacement seen within the barrier will be extremely small if it has been designed and constructed well. The striking vehicle also has elasticity and two elements are considered; elasticity between the impulse force and the vehicle's generalized mass center (K_{Sm}) and elasticity between the impulse force and the friction force occurring at the tire patch (K_{Sf}). As one would expect, the influence of K_{Sf} is ignored when the friction is ignored.

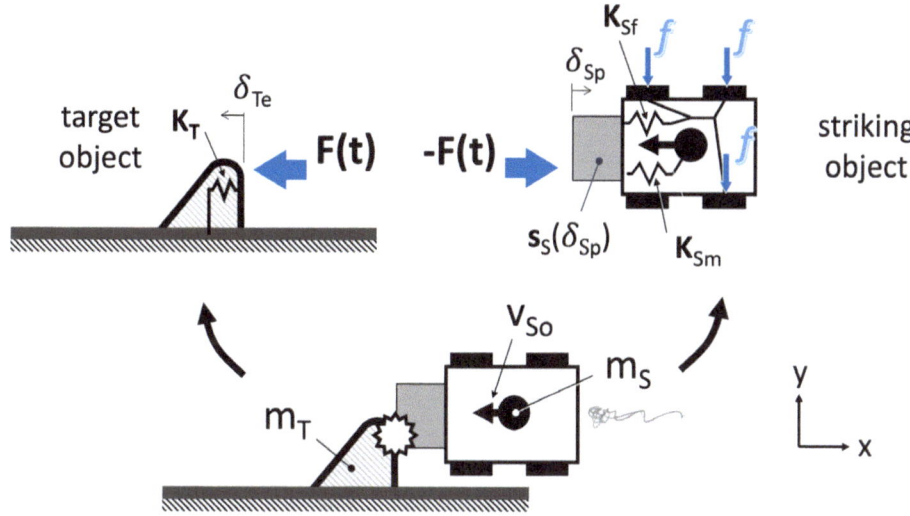

Fig. 4.46 General SOF collision model

Fig. 4.46 variable definitions:

s_S	=	is the strength profile of the striking vehicle,
δ_{Sp}	=	is the direction of striking vehicle plastic deformation measure,
δ_{Se}	=	is the direction of striking vehicle elastic deflection measure,
K_{Sm}	=	is the stiffness of the striking vehicle between impulse and mass center,
K_{Sf}	=	is the stiffness of the striking vehicle between impulse and tire patch,
δ_{Te}	=	is the direction of barrier elastic deflection measure,
KT	=	is the stiffness of the barrier

Insight from Governing Equations

We can leverage the collision model and equations to develop an understanding of how the impulse magnitude responds to the loadcase variables. We will start with the impulse (J) equations stating that impulse is equal to the momentum change of an object involved in the collision (Eq. 4.67) and equal to the area under the F·t curve (Eq. 4.68).

$$J = p_o - p_f \qquad [\text{N s] or [kg m/s}] \tag{4.67}$$

$$\text{Impulse}: \boxed{\,J = \int_o^f F(t)dt\,} \tag{4.68}$$

or

$$\boxed{J = F_{ave}\Delta t} \tag{4.69}$$

where:

J	=	impulse
p	=	momentum of a collision object
F	=	impulse force
Δt	=	time duration of collision

$$J = m_S v_{Sf} - m_S v_{So} \tag{4.70}$$

$$F_{ave}\Delta t = m_S \left(v_{Sf} - v_{So}\right) \tag{4.71}$$

where:

m_S	=	mass of the striking vehicle
v_{Sf}	=	final velocity of the striking vehicle
v_{So}	=	initial velocity of the striking vehicle

Equation 4.71 shows us that the amount of impulse applied to the striking vehicle is a function of its velocity change; the less the change in velocity, the less the impulse. This relationship is noteworthy with regards to SOF, as different vehicles can exhibit different behavior and different levels of velocity change during test.

We can also relate the amount of work done on the striking vehicle to loadcase parameters. To do this, we will exercise the conservation of energy law...

$$E_f = E_o \tag{4.72}$$

$$\left(KE_f + W\right) = KE_o \tag{4.73}$$

$$\left(\frac{1}{2}m_S v_{Sf}^2\right) + W = \left(\frac{1}{2}m_S v_{So}^2\right) \tag{4.74}$$

$$W = \frac{1}{2}m_S \left(v_{So}^2 - v_{Sf}^2\right) \tag{4.75}$$

$$\boxed{F_{ave}\Delta_{Sp} = \frac{1}{2}m_S \left(v_{So}^2 - v_{Sf}^2\right)} \tag{4.76}$$

where:

m_S	=	mass of the striking vehicle
v_{So}	=	initial velocity of the striking vehicle
v_{Sf}	=	final velocity of the striking vehicle
F_{ave}	=	average impulse force
Δ_{Sp}	=	crush distance seen within striking vehicle

Equation 4.76 shows us that that some of the striking vehicle's kinetic energy will be converted into material fracture and crush. The amount of energy transfer depends on the velocity change.

System Energy Flow
As Eq. 4.76 shows, some of the initial kinetic energy is converted into crush within striking vehicle ($F_{ave}\Delta_{Sp}$). Residual kinetic energy will exist at the end of the test, the amount of which depends on the crash mode exhibited. In cases where the residual kinetic energy is high, the work done through crush is low.

Energy will also be stored during the event, the amount of which is dependent on the striking vehicle's stiffness (K_{Sm}), the maximum elastic compression seen within the striking vehicle (Δ_{Se}), the barrier's stiffness (K_T), and the maximum elastic compression seen within the barrier (Δ_{Te}). This energy is released once the impulse force is removed. The amount of stored energy will depend on the magnitude of the impulse force that occurs during the event. Figure 4.47 illustrates the system energy flow.

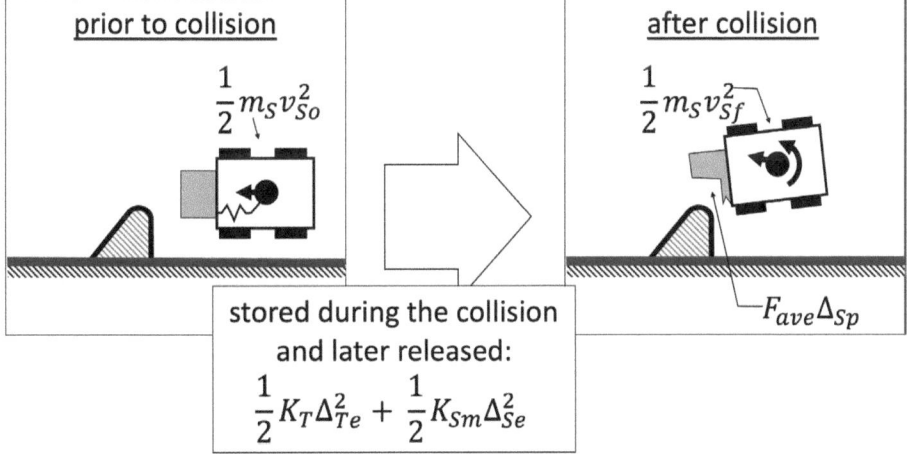

Fig. 4.47 SOF energy conservation

Crash Modes

Vehicles tested in the SOF loadcase will exhibit one of two different trajectory behaviors or a behavior somewhere in between these two. The first behavior, or crash mode, can be described as "bounce" (Fig. 4.48b, Link 4.10). Here, nearly all the vehicle's initial kinetic energy is expended as crush. The second crash mode can be called "glance" (Fig. 4.48a, Link 4.11). A vehicle exhibiting this behavior will pass by the barrier and retains a significant amount of its original kinetic energy at the end of the event. These two crash modes can be considered the extreme ends of the behavior spectrum; the behavior of any given vehicle is often somewhere in between these two conditions (Link 4.12). We will find that the implications to the structure as well as the loadpath topology is significantly different depending on where a vehicle's crash mode lays within this spectrum.

|  | Pure Bounce: SOF test of a 2014 Acura MDX (https://www.youtube.com/watch?v=WaHlmNaNZ6Q) IIHS YouTube channel | (Link-4.10) |

|  | Pure Glance: SOF test of a 2021 Tesla Model-Y (https://www.youtube.com/watch?v=5G1dF392iys) IIHS YouTube channel | (Link-4.11) |

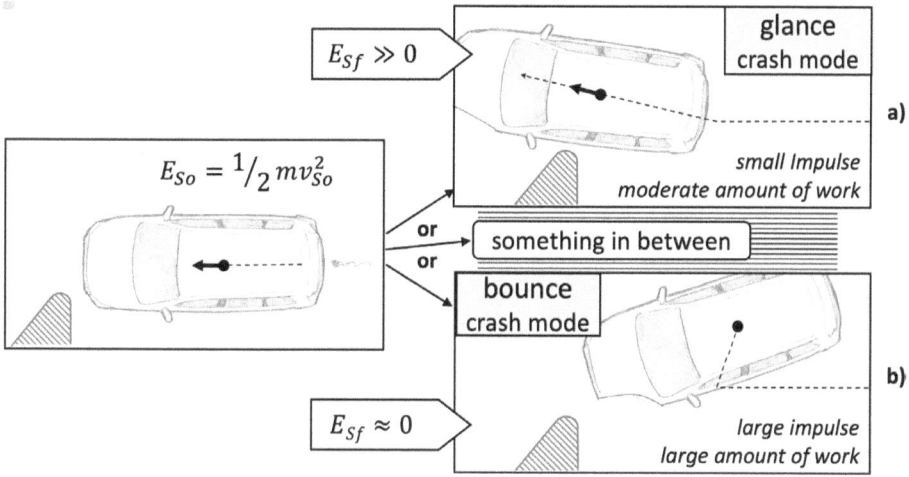

Fig. 4.48 Crash modes of SOF

	In Between: <u>SOF test of a 2012 Volvo S60</u> (https://www.youtube.com/watch?v=YSYLE55iYj0) IIHS YouTube channel	(Link-4.12)
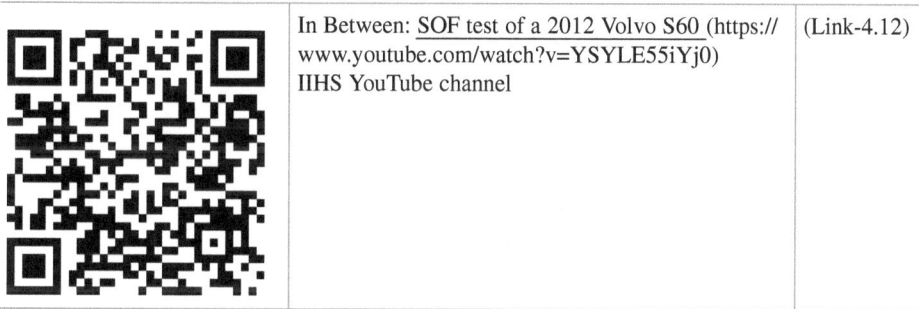		

The collision model, vehicle dynamics, and energy dissipation are quite different depending on the crash mode, so they will be covered separately, starting with the glance crash mode.

4.3.7   The Glance Crash Mode

Fundamental Behavior

A vehicle exhibiting a glance crash mode will collide with the barrier but ultimately pass by the barrier during the event, as illustrated in Fig. 4.49.

The impulse of the collision will change the velocity of the striking vehicle; reducing its magnitude and changing its direction. The FDB shown in Fig. 4.50 illustrates that

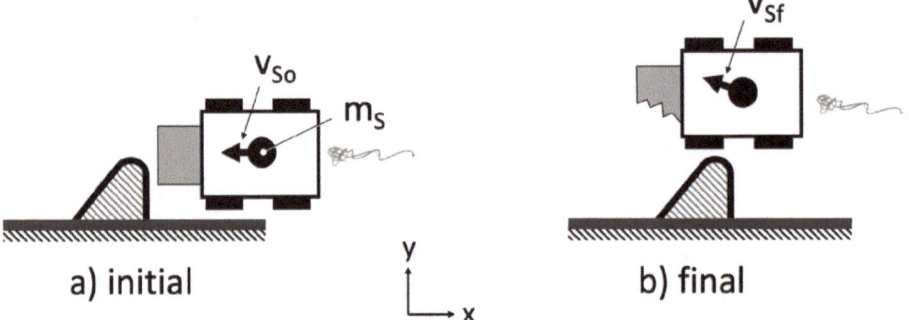

Fig. 4.49 SOF Glance crash mode collision model

the striking vehicle's momentum and friction between the tire and ground will resist the impulse force. The effect of friction (f and K_{Sf}), however, is typically ignored.

The time steps of a 'glance' crash mode are illustrated in Fig. 4.51.

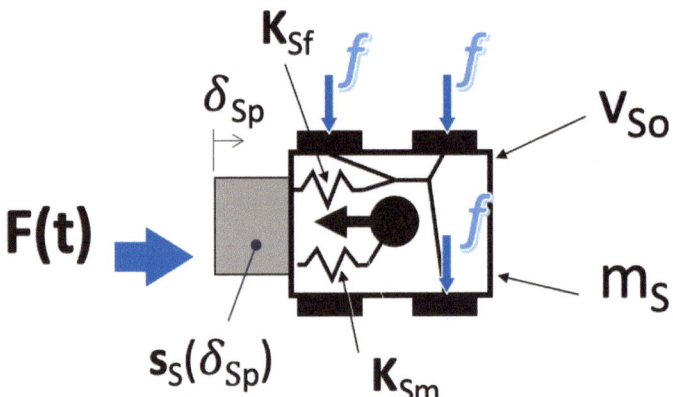

Fig. 4.50 SOF glance FBD

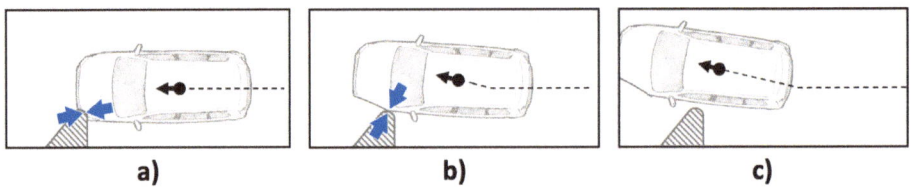

Fig. 4.51 SOF glance crash mode timestep

Governing Physics

A glance crash mode exists when the test vehicle achieves a level of lateral velocity during the collision such that the *front-body-hinge-pillar* is positioned outboard of the barrier when it reaches it (Fig. 4.51b); the vehicle will pass by the barrier without significant longitudinal impulse force between the barrier and the *front-body-hinge-pillar*.

Notice how the direction of the impulse force changes for a glancing vehicle during the event, as the vehicle translates laterally away from the barrier and interfaces the radiused edge of the barrier. Here, the impulse force during the event will have both a fore-aft and a lateral component. The lateral component is important, as building lateral velocity during the event is key to having the vehicle pass by the barrier.

The lateral components of impulse, the associated lateral force, and the vehicle's lateral velocity are illustrated in Fig. 4.52. Note that the lateral force curve is a compellation of three different $F_y(t)$ curves (dashed curves in Fig. 4.52), each representing an impulse event that occurs at a different loadpath (barrier contact with *front-impact-beam-extension*, barrier contact with *front-cradle*, etc.). Also, the magnitude of each force curve gradually increases in accordance with the strength and stiffness of each engaged loadpath. Per our collision model and Eq. 4.67, it is this lateral component of the impulse event that changes the vehicle's momentum, adding a lateral component to its velocity.

The impulse magnitude required to achieve a glance crash mode varies from vehicle to vehicle, largely dependent on the vehicle's mass, size, proportions, and suspension geometry. Since impulse is a force–time product, increasing the duration of structural engagement is important for achieving the necessary impulse magnitude and keeping the impulse force magnitudes manageable. Alternatively, imagine the strength and stiffness that would be required from a single loadpath to achieve the necessary impulse magnitude, as illustrated in Fig. 4.53. It should be no surprise that most vehicles that exhibit a glance

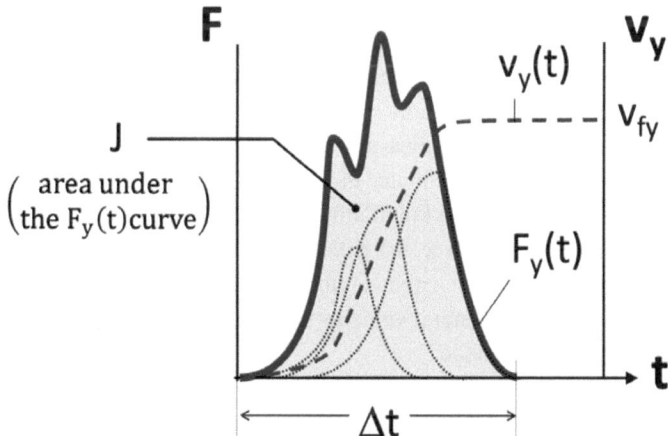

Fig. 4.52 Lateral components of impulse

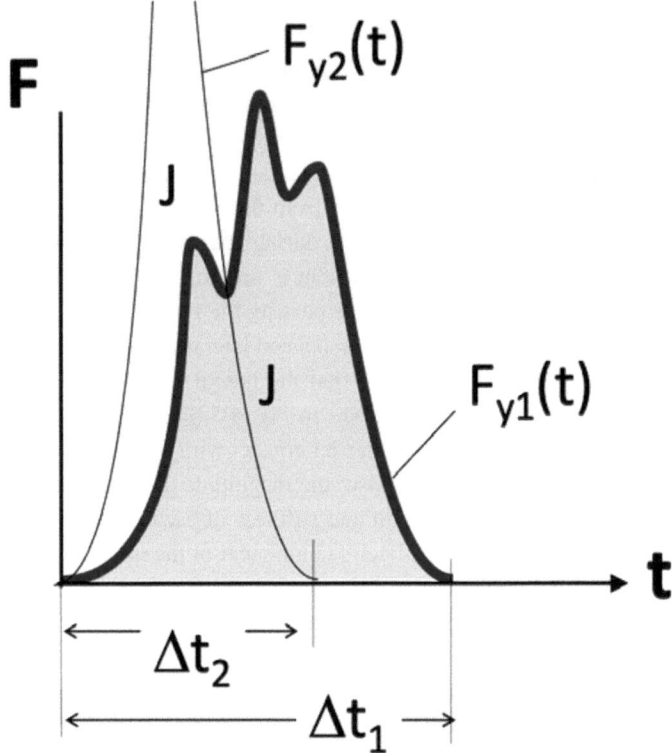

Fig. 4.53 Lateral components of impulse, a comparison

crash mode build the necessary lateral velocity by successively engaging the barrier with different loadpaths.

Force must be applied over a duration of time to achieve the impulse magnitude required to sufficiently change the object's velocity. This is another key concept to absorb. Recognize that, although the duration can be extremely small, the force between two collision objects occurs over some measurable duration of time.

With respect to a glance crash mode, Fig. 4.54 illustrates that there is a relationship between the lateral impulse and the longitudinal distance between the barrier and the *front-body-hinge-pillar*. In order to place the *front-body-hinge-pillar* outboard of the barrier, vehicles with a smaller d_x must engineer the vehicle to achieve the necessary lateral impulse quicker. In practice, a lateral velocity(t) target is typically identified for a vehicle designing for a glance crash mode.

Recall the lesson of Sect. 3.2.2; a force directed at an object's mass is required to change its velocity. This is a key concept for understanding the glance crash mode; there must be an effective loadpath between the impulse force and the striking vehicle's mass concentration if its direction is to be changed.

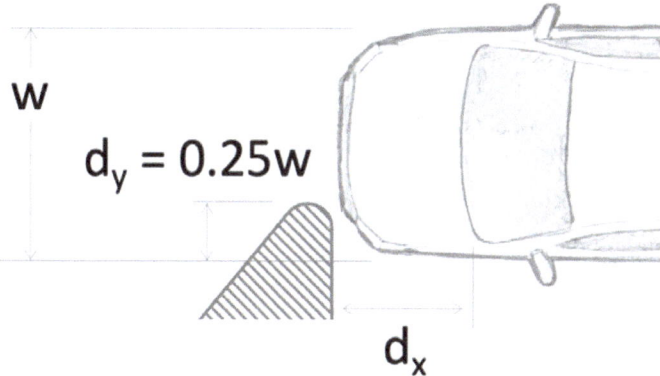

Fig. 4.54 Influence of front-body-hinge-pillar location

The mass concentration of a typical BFI ICE automobile is illustrated in Fig. 4.55. A powertrain in the front compartment comprises a meaningful portion of the vehicle mass. Consider that the rest of the vehicle's significant mass concentrations are rearward of the *dash-panel.*

We can assume the masses rear of the *dash-panel* to be lumped, as the vehicle structure in this region is relatively stiff with exceptional plan-view shear stiffness (Fig. 4.56b). The front-compartment structure, however, has limited plan-view shear stiffness, primarily because of the spatial requirements of front compartment content (powertrain, etc.) and the crushability characteristics required for the Full-Frontal and MOF loadcases (Fig. 4.56a).

The front-compartment structure of a typical BFI ICE automobile can be considered a ladder frame construction, relatively soft in plan-view shear stiffness. Application of lateral impulse forces to the front of a soft front-compartment structure matchboxes the front structure and does little to instigate lateral acceleration of the concentrated vehicle mass behind the *dash-panel*, as illustrated in Fig. 4.57b.

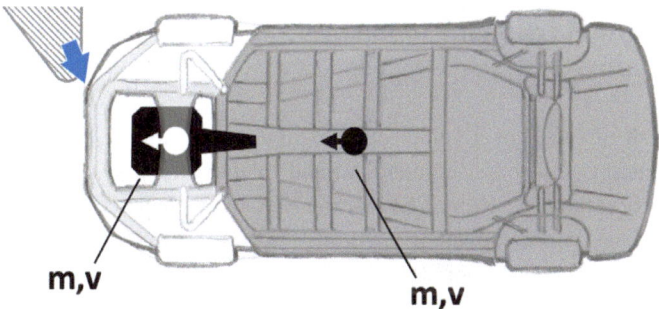

Fig. 4.55 Typical BFI ICE mass concentrations

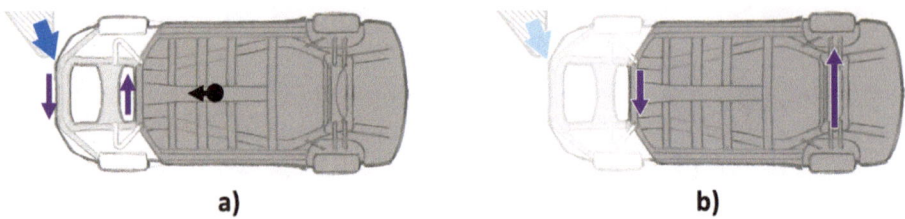

Fig. 4.56 Typical BFI ICE plan-view shear stiffness

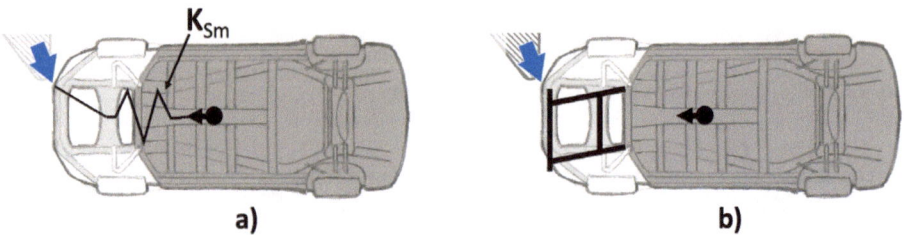

Fig. 4.57 Behavior of a front compartment structure soft in plan-view shear

Recall the 'pushing a springy object against a wall' experiment described in Sect. 3.2.3.2. The effectiveness of the front compartment with regards to transmitting force to the vehicle mass behind the *dash-panel* is dependent on its stiffness; it is part of the elasticity loadpath between the impulse force and the mass that needs to be accelerated (K_{Sm}), as illustrated in Fig. 4.57a.

In short, to achieve a glance crash mode;

- lateral force must be applied to the vehicle mass concentrations, and
- the loadpath between the applied force and the mass concentrations needs to be stiff to be effective.

4.3.7.1 Structural Loadpath Topology

Figure 4.58 illustrates the typical loadpath topology of a vehicle exhibiting a glance crash model in SOF.

The loadpath topology of a vehicle that exhibits a true glance mode will utilize relatively little energy absorption content; limited to the very first structural elements to contact the barrier; the *front-impact-beam-extension* and elements of the *front-cradle*. The rest of the topology will be aimed at providing stiffness between the applied barrier load and the lumped vehicle mass behind the dash. Anything that increases the plan-view shear stiffness of the front-compartment is within scope of this portion of SOF glance mode topology and there are many effective design elements that can be employed. Recall that

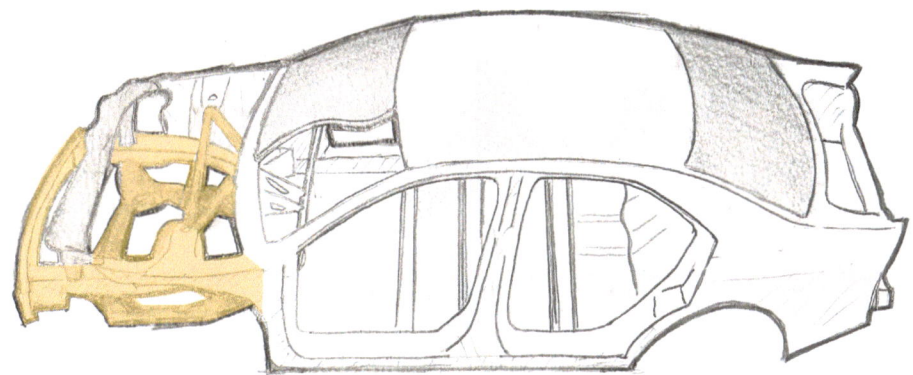

Fig. 4.58 Loadpath topology of an SOF glance crash mode

executing a glance crash mode typically requires multiple, successive loadpaths. Hence, more than one of the elements illustrated in Fig. 4.59 are usually found in a vehicle that glances off the barrier.

a. **"Splaying" the body** *front-compartment-mid-rail* such that its barrier engagement is improved.
b. **Extending the** *front-impact-beam* **outboard** (*front-impact-beam-extension*) and,
c. **including a "blocker" or "wedge"** attached to the rear side of the *front-impact-beam-extension*. The *beam-extension* will fold rearward upon contact with the barrier, as illustrated in Fig. 4.60. The amount that the *beam* is extended can be strategically driven based on a targeted fore-aft location for contact on the *front-compartment-mid-rail* (q). It is desirable that the contact location on the *mid-rail* be a point which is stiff laterally; near a body crossmember or a front cradle interface location (x), for example. *-note that the effectiveness of these elements is highly dependent on the plan-view shear stiffness of the front compartment; boosting that stiffness is often a prerequisite for these elements to be effective.*
d. **positioning a body crossmember** connecting the *front-compartment-mid-rails*. Effectiveness can be sensitive to the crossmember to *mid-rail* joint stiffness about the z-axis; as plan-view shear stiffness is critical.
e. **gusseting the front-compartment-mid-rail to the** *1-bar* to improve the joint stiffness about the z-axis at this location; and increasing the plan-view shear stiffness of the front compartment.

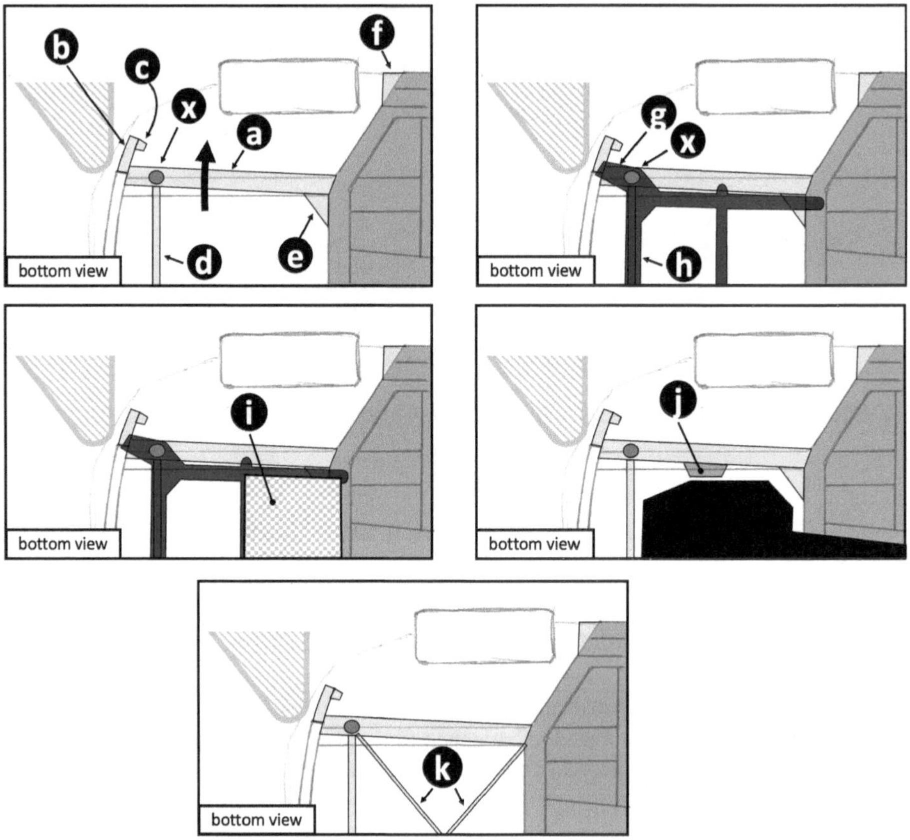

Fig. 4.59 SOF Glance crash mode design elements

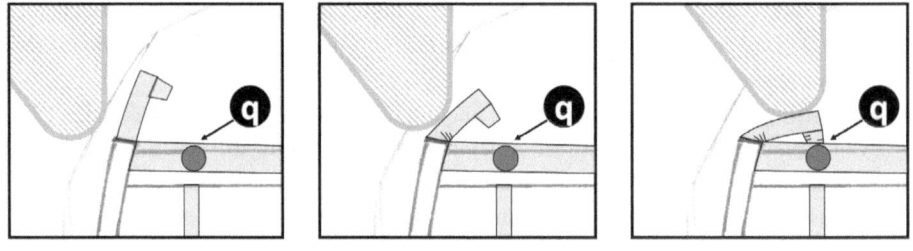

Fig. 4.60 Front-impact-beam-extension and blocker behavior

f. **chamfering the leading edge of the front-body-hinge-pillar** to promote the vehicle shedding off the barrier upon impact at this location.

g. **incorporating *front-cradle* elements to engage the barrier**. This popular design element introduces the strength and stiffness of the *front-cradle* into the impulse event.

h. **positioning the *front cradle front-crossmember* in line with body attachment location** (x). This maximizes the effectiveness of the *front-crossmember* in stiffening the front compartment in plan-view shear.

i. **introducing shear-panels.** Locations for shear-panels must consider the crush zones and crush strategies of other loadcases.

j. **engineering a direct or more direct loadpath to powertrain.** Some vehicle designs incorporate a "blocker" that is packaged on the inboard side of the *front-compartment-mid-rail* such that the powertrain is contacted by the *rail* when it deflects inboard during the event. Although clearance between the *blocker* and the powertrain exists to ensure there is no contact during normal vehicle usage, the clearance is minimized so that the powertrain can be accelerated as soon as possible. The *blocker* is effective because its loadpath to the powertrain is more direct and stiffer than the existing loadpaths that travel through the elastomeric powertrain mounts.

k. incorporating an **x-brace** into the front-compartment structure to increase the plan-view shear stiffness. Connecting the body *mid-rails* with an x-brace is effective when the *mid-rails* are engaged by the barrier. Similarly, an x-brace incorporated into the *front c*radle is effective when it is engaged by the barrier. Incorporating an x-brace can be difficult due to the spatial requirements of front compartment content (powertrain, etc.) and crush/deformation characteristics required for other front crashworthiness loadcases.

Specific vehicle solution examples are shown and described later in this chapter.

Perhaps everything described in the 'glance' portion of this section sounds like 'designing to the test', meaning that the solution is engineered to perform well on a standardized test but not necessarily in real world crash events. This is completely understandable and the skepticism is healthy. Although there is little data comparing performance on the SOF barrier and real word crashes, performance correlation is indicated. One example is from testing performed during the IIHS SOF protocol development. Here, the 2009 Mitsubishi Galant demonstrated glance crash mode characteristics in both the SOF barrier and car-to-car testing. Dummy injury metrics from the Galants were also similar between these two test conditions (IIHS tests; CF11003, CF11010, CF11010).

4.3.8 The Bounce Crash Mode

Fundamental Behavior

The vehicle crashes straight into the barrier in a true bounce crash mode.

Work is done through crush and the vehicle undergoes a significant momentum change; starting the event with a forward velocity and ending the event at rest. Figure 4.61 illustrates the system with respect to conservation of momentum.

Time steps of a 'bounce' crash mode are illustrated in Fig. 4.62. The direction of the impulse force changes little during the event, remaining largely in the x-axis. Near the end of the event, the vehicle springs off the barrier due to elasticity in the system and the vehicle rotates in top view due to the lateral offset between the vehicle cg and the barrier face/ impulse force. Frictional forces acting on the tires bring the striking vehicle to a rest.

A vehicle exhibiting a bounce crash mode undergoes more momentum change than a glancing vehicle of similar mass.

Governing Physics

The physics of an SOF bounce crash mode is very similar to other crash loadcases where the test vehicle's final velocity is zero; the impulse force and the crush must be managed such that the structural intrusion and dummy metrics are satisfactory. Impulse loading

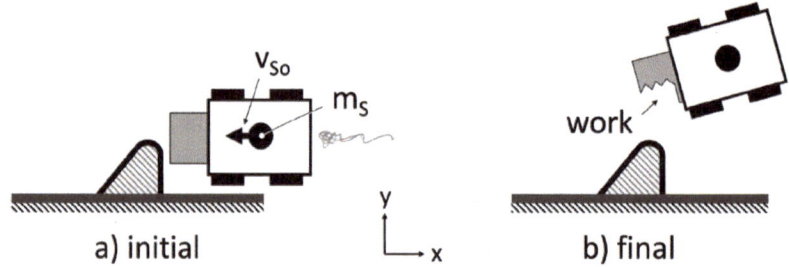

Fig. 4.61 SOF Bounce crash mode collision model

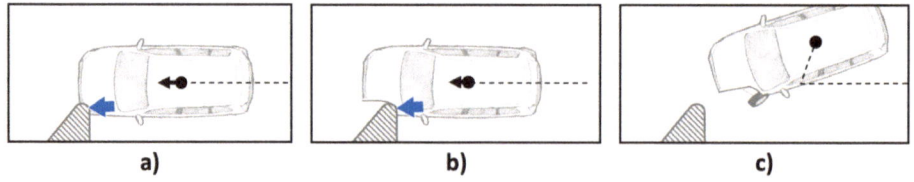

Fig. 4.62 SOF bounce crash mode timestep

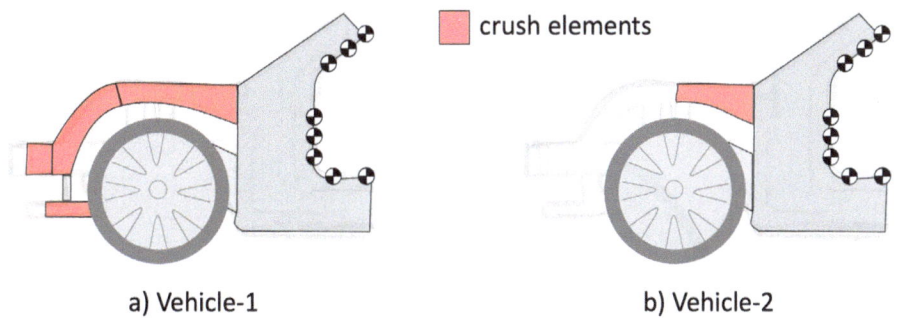

a) Vehicle-1   b) Vehicle-2

Fig. 4.63 Energy absorption elements; quantity and location

occurs on the structure from direct barrier contact and from the wheel/tire, which is typically captured between the barrier and the structure in a bounce crash mode.

The key to an efficient design is to engineer energy absorbing crush elements that performing work across as much of the impulse time duration as possible. Broadening the crush participation in this way minimizes force spikes seen by the back-up structure and reduces the strength demanded of the back-up structure; recall that impulse is a F·t quantity.

Consider two theoretical vehicles; both exhibiting a bounce crash mode and both with equivalent post-test door opening measurements, as illustrated in Fig. 4.63. Vehicle-1 engages crush elements through a large time duration while Vehicle-2 has a much smaller time duration of crush element engagement.

For Vehicle-2 to perform similarly to Vehicle-1, it must manage a similar longitudinal impulse (area under the F·t curve) and its back-up structure must manage a much higher force without yielding; $F_{x2}(t)$ versus $F_{x1}(t)$ in Fig. 4.64.

As with other front crash loadcases, crush elements must be designed such their strength is less than that of the back-up structure to ensure that plastic deformation occurs forward of the occupant compartment. Specifically, strength of the zone-A (S_A) and strength of zone-B (S_B) must be lower than the strength of the back-up structure (S_{BUS}), as illustrated in Fig. 4.65.

4.3.8.1 Structural Loadpath Topology

As the physics would suggest, the loadpath topology of a vehicle with a bounce crash mode encompasses the near-side of the vehicle structure (as illustrated in Fig. 4.66) and the structural design elements are intended to increase the presence of crushable energy absorbing elements and to strengthen the back-up structure.

Such a vehicle likely has many of the design elements outlined in Fig. 4.67.

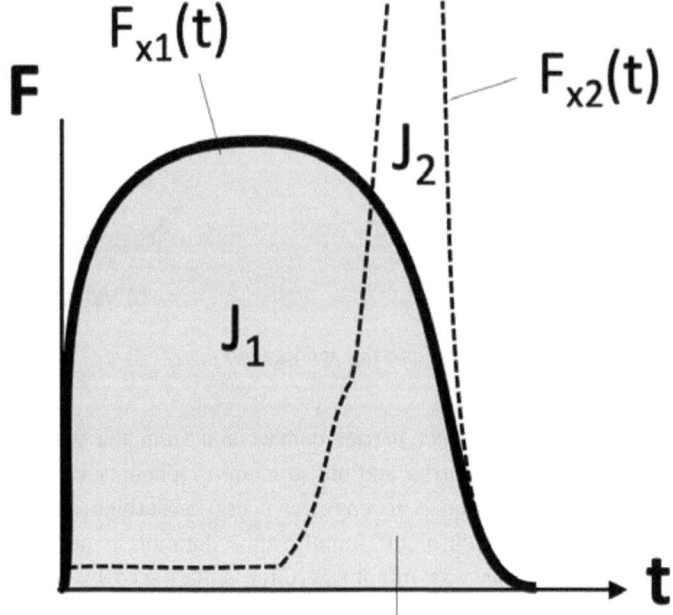

Fig. 4.64 Longitudinal components of impulse, a comparison

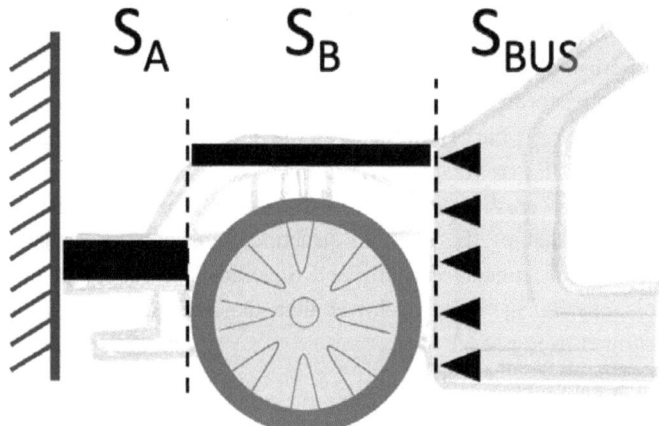

Fig. 4.65 Loadpath strength allocation

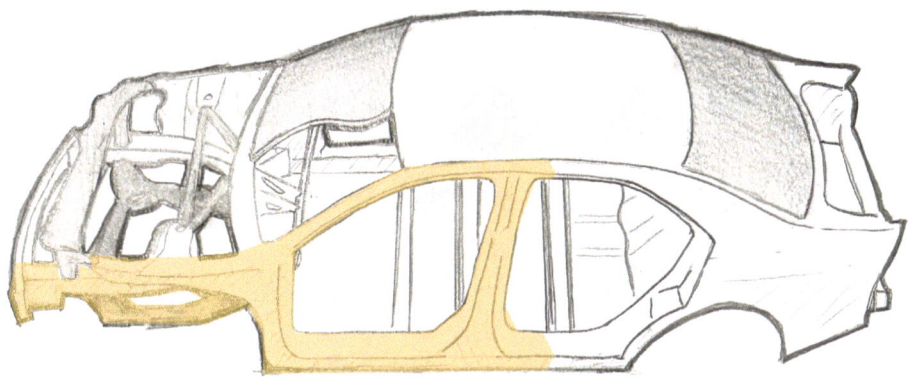

Fig. 4.66   Loadpath topology of an SOF bounce crash mode

(a) **extending the *front impact-beam crush-can*** outboard such that it engages the SOF barrier. Note that this design element will affect the crush behavior in other front crash loadcases and some level of design tuning will be required.

(b) **Strengthening or adding an 'elephant trunk'**; a member connecting *front-compartment-upper-rail* to the widened crush-can and/or the front of the *front-compartment-mid-rail*.

(c) **strengthening the *front-compartment-upper-rail***

(d) **strengthening the *front-body-hinge-pillar-upper* and *A-pillar*** to manage the upper loadpath crush forces and act as a non-deforming back-up structure

(e) **incorporating a lateral extension of the *front-cradle*** such that the *front-cradle-rails* can participate in the longitudinal loadpath.

(f) **adding a connection between the *cradle, elephant trunk*, and body *mid-rail*.** This connection provides multiple loadpaths for the barrier loads applied to the widened *crush-can* and eventually to the *elephant-trunk* itself.

(g) **tuning the *wheel* strength**

(h) **adding body *1-bar-extension* reinforcement** to manage direct loading of the wheel/tire.

(i) **strengthening the lower portion of the *front-body-hinge-pillar*, *A-pillar*, and *rocker*** to manage the lower loadpath crush forces and act as a non-deforming back-up structure.

(j) **strengthening the *bodyside-door-frame*** to act as an unyielding back-up structure for the crush forces.

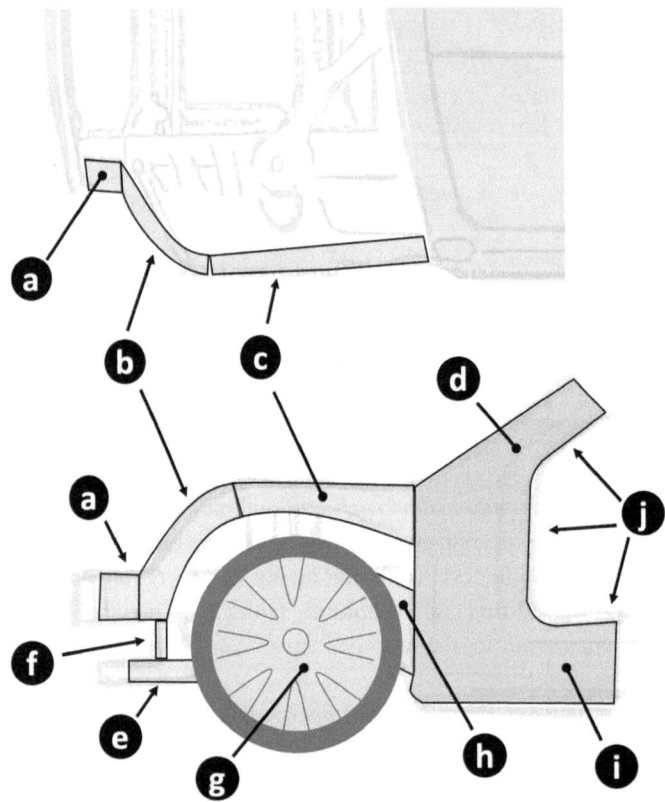

Fig. 4.67 SOF Bounce design elements

4.3.9   Wheel/Tire Behavior

Behavior of the *near-side wheel/tire* is a significant factor in SOF performance and requires consideration in a vehicle's crashworthiness development. The *tire* deflates during the event but the *wheel* remains a relatively stiff and strong element to contend with. Recall the rating measurement points illustrated in Fig. 4.44; it is undesirable for the wheel to cause significant deformation to the *front-body-hinge-pillar*, side-door opening and/or the footwell.

Kinematics
The *wheel/tire* will stop moving once it contacts the barrier, however the rest of the vehicle will continue to move forward at this point, regardless of the vehicle's crash mode. This creates a relative velocity difference between the *wheel/tire* and the rest of the vehicle, as illustrated in Figs. 4.68b and 4.69b. Kinematics of the *wheel/tire* are now primarily dictated by the linkages which connect it to the vehicle. Specifically, it is the geometry

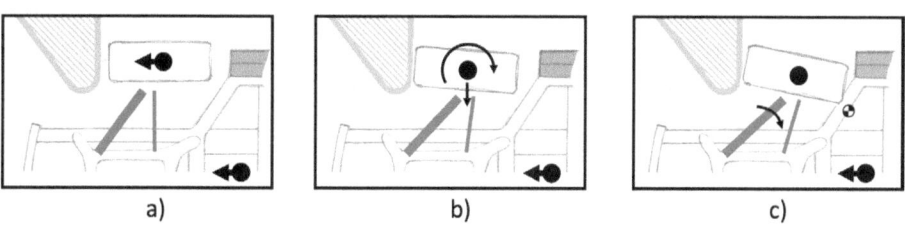

Fig. 4.68   Wheel/tire kinematics with a forward facing lower-control-arm (bottom view)

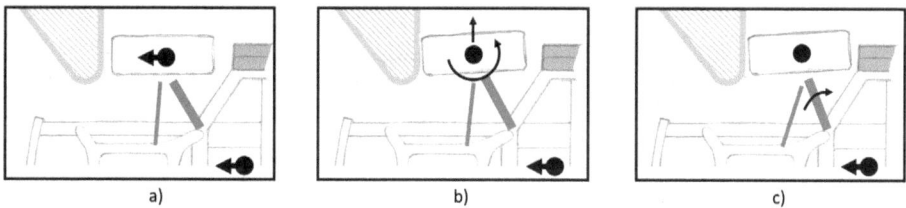

Fig. 4.69   Wheel/tire kinematics with a rear facing lower-control-arm (bottom view)

and orientation of the front suspension *lower-control-arm* that are important. Figures 4.68 and 4.69 illustrate the two typical configurations of the *lower-control-arm*; a forward facing configuration (Fig. 4.68) and a rear-facing configuration (Fig. 4.69). These figures illustrate in timestep the wheel/tire kinematics for the two configurations.

Forward Facing Lower-Control-Arm

Figure 4.68 shows that *wheel/tire* kinematics of a vehicle with a forward facing *lower-control-arm* configuration is dictated by the articulation of the forward link. In this configuration, the *wheel/tire* rotates inward and is drawn inboard during the SOF event. Note that the translated *wheel/tire* does not have good alignment to the *rocker* and contacts the relatively weak *1 bar-extension/dash* area. Vehicles in this condition tend to incorporate strong reinforcements between the *front-compartment-mid-rail* and the *rocker* to manage the forces and reduce dash intrusion.

Rear Facing Lower-Control-Arm

Figure 4.69 shows that the *wheel/tire* kinematics of a vehicle with a rear facing *lower-control-arm* configuration is dictated by the articulation of the rear link. Here, the wheel/tire rotates outward and translates outboard. The kinematics are more favorable here, aligning the wheel to the rocker and positioning it further outboard.

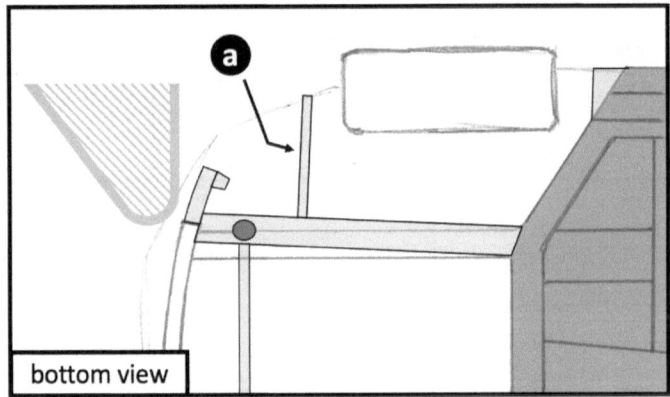

Fig. 4.70 Design element to influence wheel/tire kinematics

Influencing Wheel/Tire Kinematics

The ability of structure to control the wheel kinematics is quite limited, however there is one notable design element that exists.

a. Introducing a **wheel/tire rotation arrestor** can be beneficial in cases where the *wheel/ tire* rotation behavior is unfavorable ('a' in Fig. 4.70). This can be a newly added component to the structure or the function can be incorporated into an existing component like an outboard powertrain cooler bracket. Upon contact with the barrier, the arrestor folds about its connection at the body *front-compartment-mid-rail* upon barrier contact, rotates rearward, and eventually contacts the front of the tire. Although the suspension geometry will tend to drive the position of the wheel, this countermeasure will reduce *wheel/tire* rotation once the arrestor contacts the tire.

Wheel/Tire Separation and Ejection

Separating the *wheel/tire* from the vehicle and ejecting it from the wheel opening is advantageous because it can reduce the amount of loading that occurs on the *front-body-hinge-pillar*. As we will see, separation only occurs in vehicles that exhibit a significant amount 'glance' in the crash mode.

Consider the illustrations shown in Figs. 4.71 and 4.72. Forward motion of the *wheel/ tire* is prevented once it contacts the barrier, while the rest of the vehicle maintains some level of forward velocity. As the vehicle continues to move forward, the linkages connecting the wheel to the rest of the vehicle will articulate, undergo tensile loading, and eventually fracture, as illustrated in the third frames. Notice how tensile forces in the rear-link of the *lower-control-arm* occur later than in the front-link. If the rear-link is to fracture, the vehicle must still have opposing velocity at this stage of the event. This is not the case in a pure bounce crash mode, as illustrated in the time frames of Fig. 4.72.

We now know that sufficient tensile forces must be generated in every connecting link for the *wheel/tire* to separate from the rest of the vehicle and that relative velocity is required to generate those tensile forces. To achieve *wheel/tire* separation, **deflection must be sufficient for the inboard attachment location of each linkage** (Fig. 4.73) **to pass by the barrier**, as there will be no relative velocity otherwise. Figure 4.74 illustrates the condition where the *spring/damper* attachment passes by the barrier while Fig. 4.75 illustrates the condition where it does not.

The IIHS-SOF test of a 2015 Volkswagen GTI illustrates the condition where the *spring/damper* attachment does not pass by the barrier and the strut linkage is not fractured during the event (Link 4.13).

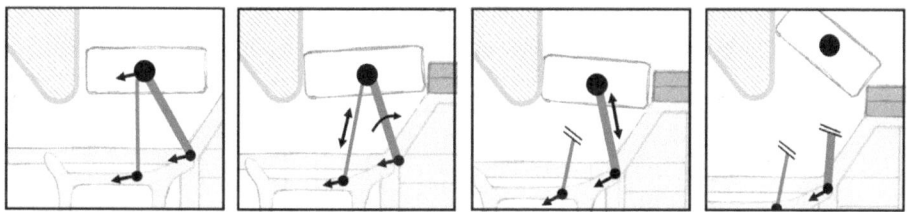

Fig. 4.71 SOF glance crash mode time steps: Lower-control-arm and wheel/tire behavior (bottom view)

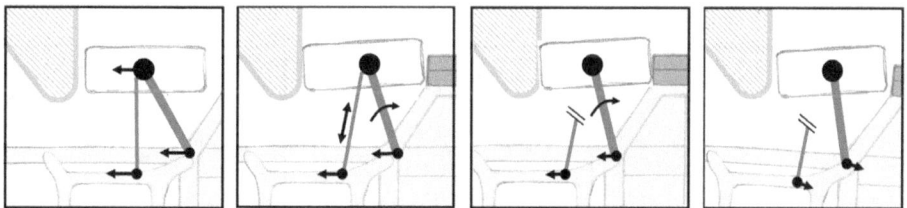

Fig. 4.72 SOF bounce crash mode time steps: Lower-control-arm and wheel/tire behavior (bottom view)

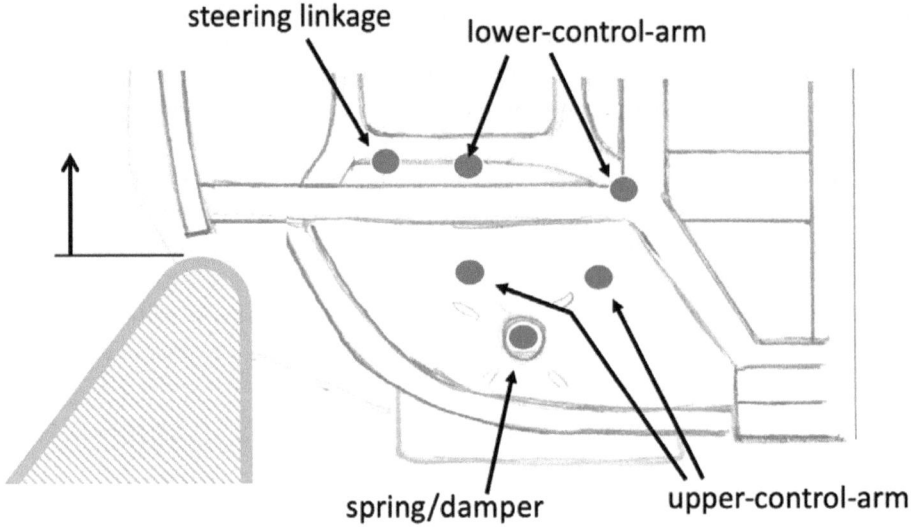

Fig. 4.73 Typical inboard attachment locations of front suspension linkages

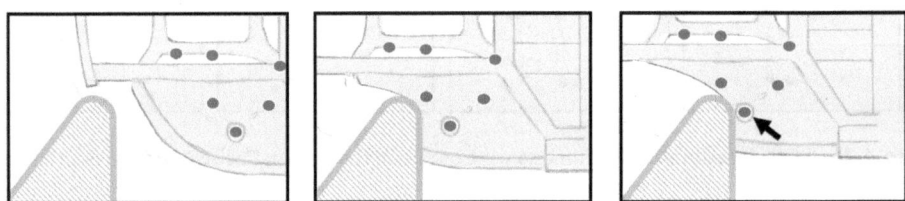

Fig. 4.74 All inboard linkages pass by the barrier

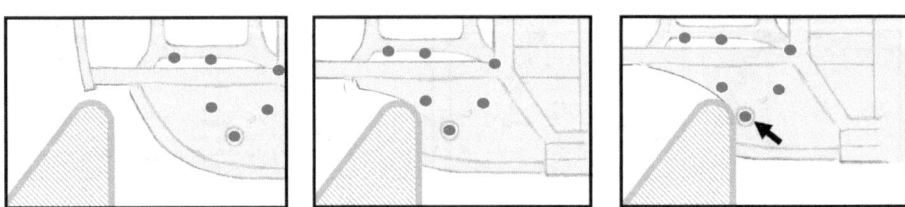

Fig. 4.75 Spring/damper inboard attachment 'hang-up' on the barrier

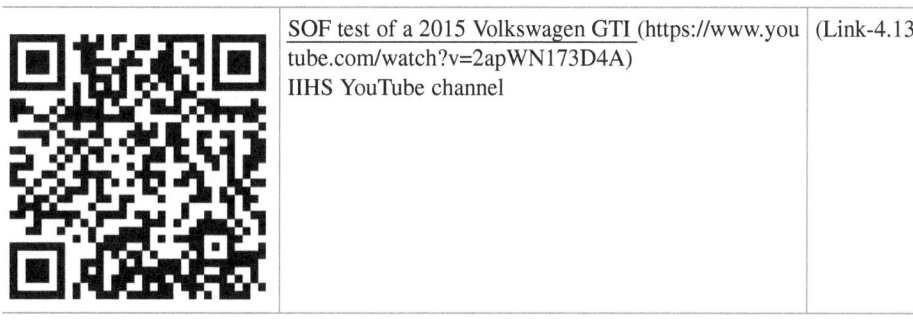

SOF test of a 2015 Volkswagen GTI (https://www.you tube.com/watch?v=2apWN173D4A) IIHS YouTube channel | (Link-4.13)

In addition to each inboard attachment location of each linkage passing by the barrier, **the forward velocity at the time each inboard attachment passes by must be sufficient to generate the necessary tensile force**. It is entirely possible that all inboard attachments pass by the barrier but a component of bounce crash mode reduces the relative motion enough that the wheel/tire is not separated. The video of a 2013 Chevrolet Cruze illustrates this (Link 4.14).

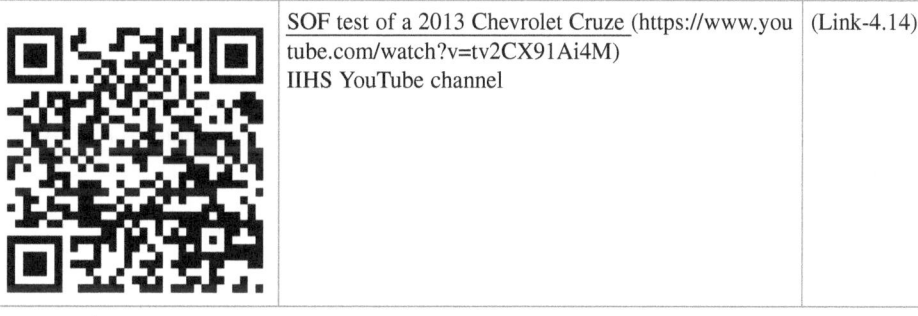

SOF test of a 2013 Chevrolet Cruze (https://www.you tube.com/watch?v=tv2CX91Ai4M) IIHS YouTube channel | (Link-4.14)

To be fair, there are other factors that influence separation, including; the strength of the linkage attachment, the stiffness and strength of the structure at the inboard attachment, etc.

Wheel Rim Position
The lateral position of the wheel lip relative to the *front-body-hinge-pillar* is also an important variable in wheel expulsion. If the lip is positioned outboard the *front-body-hinge-pillar*, the *wheel* will tend to fracture and the *wheel* will be expelled. Alternatively, if the lip is inboard of the *front-body-hinge-pillar*, the relatively strong outside face of the rim tends to get stuck in the wheel opening and can result in localize loading on the body structure. Figure 4.76 illustrates these conditions.

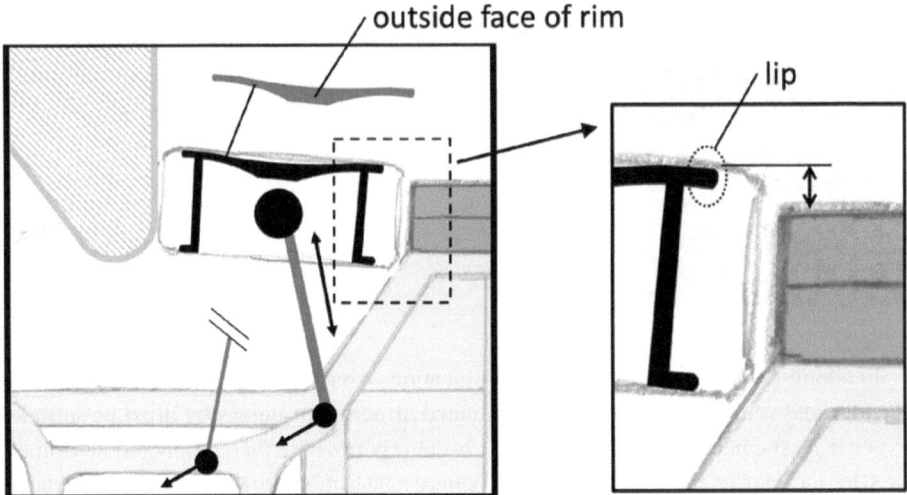

Fig. 4.76 Wheel lip position

You will recall from earlier in this section that the position and angle of the wheel/ tire's rear side during an SOF event is dictated by the suspension geometry and associated linkage kinematics, illustrated in Figs. 4.71 and 4.72. Therefore, a suspension configuration that includes a rear facing lower-control-arm is an additional enabler for wheel/tire expulsion.

4.3.10 Solution Examples

2021 Tesla Model-Y (Glance)

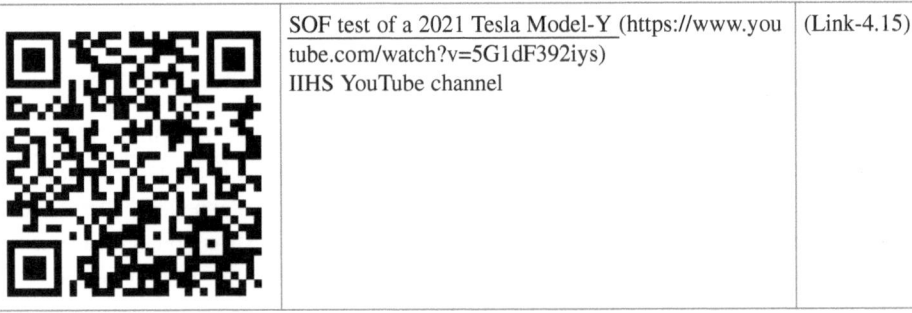

	SOF test of a 2021 Tesla Model-Y (https://www.you tube.com/watch?v=5G1dF392iys) IIHS YouTube channel	(Link-4.15)

The 2021 Tesla Model-Y tested by the IIHS demonstrated a pure glance crash mode. The vehicle deflected such that the *front-body-hinge-pillar* passed easily by the barrier and it is difficult to visually identify a reduction in forward velocity. The front, near-side *tire/wheel* separated from the vehicle and was ejected from the wheelhouse.

The body structure topology of the Model-Y is rather conventional, with splayed *front-compartment-mid-rails* and a *1-bar-upr*, as seen in Fig. 4.77. The *front-impact-beam* extends outboard of the mid-rails (a) and contacts the SOF barrier. The *beam-extension* folds and contacts the *mid-rail* during the event, as illustrated in Fig. 4.60. The topology of the *front-cradle* is somewhat unconventional, designed around a primary structure that has an arched shape, as viewed from the top or bottom. Forward extensions (b) are positioned outboard of the body *mid-rails* and contact the SOF barrier. These forward *extensions* are supported by the body *mid-rail* through an attachment at location (c) and a loadpath to the far-side body *mid-rail* through a *cradle front crossmember* (d) and the far-side body attachment (c). The connections between the *front-cradle* and body at location (c) allows the body *mid-rails* to leverage the plan-view shear stiffness of the *front-cradle* and boost the shear stiffness of the overall front compartment.

There is a loadpath from the forward extensions (b) to the vehicle's mass concentration rear of the dash through a fore-aft/diagonal outrigger that interfaces the arched primary structure of the *cradle*, at (e). The primary structure of the *cradle* attaches to the body near the bottom of the *front-body-hinge-pilar* (f) and is the primary loadpath to the vehicle's mass concentration rear of the *dash*. A mid-mount connection between the cradle and body *mid-rail* at (g) provides some shear stabilization to the arched *front-cradle.*

Although electric drive motors tend to be significantly lighter than their ICE powertrain counterparts, the Model-Y has a loadpath to the front-motor, as the front lower motor-mount attaches to the arched cradle structure at (h). Interestingly, the mounting strategy of the front-motor enables the motor to, some extent, bolster the shear stiffness of the front compartment, as the upper rear mounts are positioned at the intersection of the body *mid-rail* and *1 bar-upper* (j).

Fig. 4.77 SOF loadpath topology and elements of the 2021 Tesla model-Y

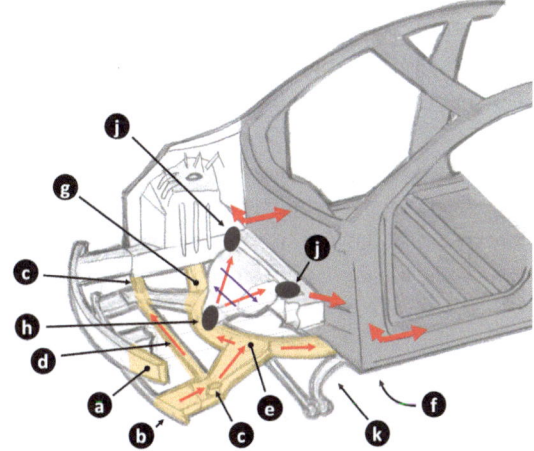

The front-suspension configuration utilizes a rear facing *lower-control-arm* (k), which is conducive to *tire/wheel* kinematics and expulsion.

2017 BMX i3, US Variant (Glance)

| SOF test of a 2017 BMW i3 (https://www.youtube. com/watch?v=l6BKlWD24YQ) IIHS YouTube channel | (Link-4.16) |

The 2017 i3 tested by the IIHS demonstrated a pure glance crash mode despite having very unfavorable vehicle proportions (the distance from the 'front of car' to the base of *A-pillar/front-body-hinge-pillar* is rather small) and having a foreword facing front suspension *lower-control-arm* configuration. Like the Model-Y, the *front-body-hinge-pillar* passes easily by the barrier and it is hard to detect a reduction in forward velocity.

Since the distance between the 'front of car' and the front of the *front-body-hinge-pillar* is rather small, the i3 must be moved laterally in a very short amount of time to achieve a glance crash mode. Perhaps not surprisingly, we will see that a highly efficient loadpath topology was required to provide such efficient vehicle dynamics.

The i3's SOF loadpath strategy is illustrated in Fig. 4.78 and starts with *front-impact-beam-extensions* (a), which folds upon contact with the barrier. The folding *impact-beam-extensions* quickly contact an angled *rail-extension*, or 'horn' (c). A claw-like feature on the back side of the *impact-beam-extension* (b) ensures that engagement onto the horn remains stable, without slip between the two elements. The *horn* transfers the impulse force into the *mid-rail* just behind the crush rails. An *x-brace* (d) connects the four corners of the *mid-rails* and significantly bolsters the plan-view shear stiffness of the front compartment, allowing effective force transfer to the mass concentration of the vehicle.

The *wheel/tire* of the i3 was ejected from the wheelhouse despite its unfavorable suspension configuration and kinematics. Separation and expulsion are enabled here because the tire is uncommonly narrow and its initial position relative to the *front-body-hinge-pillar* is significantly outboard.

Fig. 4.78 SOF loadpath topology and elements of the 2017 BMW i3

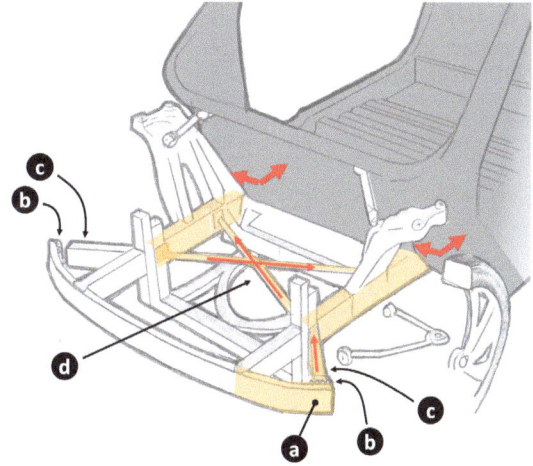

2014 Acura MDX (bounce)

	SOF test of a 2014 Acura MDX (https://www.you tube.com/watch?v=WaHlmNaNZ6Q) IIHS YouTube channel	(Link-4.17)

The MDX tested by the IIHS exhibits little lateral translation during its collision with the barrier, the *wheel* fractures and stays connected to the vehicle, and the back-up structure experiences a significant longitudinal impulse event as it contacts the barrier.

The MDX's loadpath is illustrated in Fig. 4.79. The design elements start with an extended *front-impact-beam* (a) and a body structure element positioned outboard of the *front-compartment-mid-rail* (b) which is supported by both *front cradle* structure (c) and an *elephant trunk* (d). A component connecting the front of the *elephant trunk* and the *impact-beam* acts as a tension member, pulling on the *impact-beam* and *mid-rail*, as the *elephant trunk* is crushed and moves rearward. A member connecting the top front of the *elephant trunk* to the *upper-tie-bar* (f) performs a similar function at a higher position in the vehicle. The *wheel* of the MDX fractures during the event and its material properties could have been tuned to maximize energy absorption. The *front-compartment-upper-rail*

Fig. 4.79 SOF loadpath topology and elements of the 2014 Acura MDX

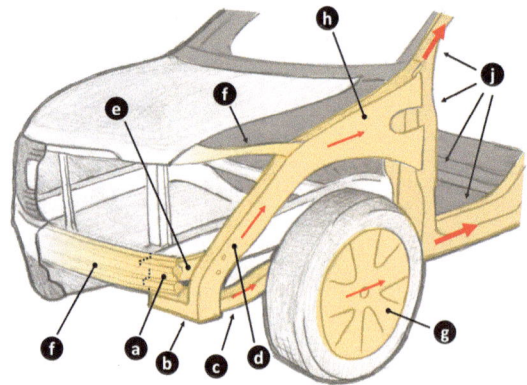

(h) crushes as it contacts the barrier. Finally, the *bodyside door-ring* (j) is constructed of ultra-high strength martensitic steel.

4.3.11 Other Considerations for SOF

Mass

The effect of mass can be described by our collision model and Eq. 4.71, wherein a heavier object will require more impulse to change its momentum. Since impulse is a F·t product, a heavier test vehicle will either need to manage higher forces, increase the impulse time duration, or some combination of both.

Consider the effect of mass on a specific vehicle (hold vehicle size, proportions, suspension geometry, loadpaths, and all other things constant) intending to execute a glance crash mode. We know that the heavier vehicle must manage a larger impulse magnitude, however the time duration of the impulse event cannot necessarily be lengthened without repercussions to structural and occupant performance.

Lengthening the impulse time duration would spread the vehicle's lateral translation over a longer time period of time, placing the inboard linkage attachments further outboard, and putting wheel/tire separation behavior at risk, as illustrated in Fig. 4.80.

Lengthening the impulse time duration would also increase the barrier to front-body-hinge-pillar overlap at t_{FBHP} and introduce a component of bounce into the crash mode, also illustrated by Fig. 4.81.

If the heavier vehicle must maintain a similar impulse time duration, Eq. 4.71 tells us that the impulse force will increase, illustrated in Fig. 4.81a. This vehicle must add glance design elements or improve the effectiveness of existing elements.

If the heavier vehicle intends to exhibit a bounce crash mode, lengthening the impulse time duration means that the length of crush would be increased. Considering that crashworthiness performance is based on displacements at the side-door-opening, increasing

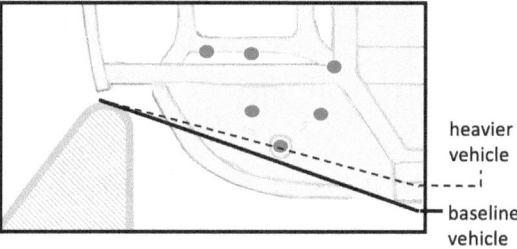

Fig. 4.80 Effect of increasing the impulse time duration, glance crash mode

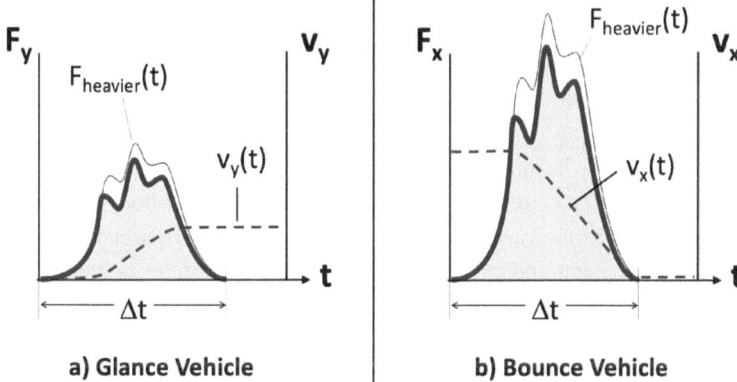

Fig. 4.81 Effect of mass on impulse

the length of crush is unacceptable. Therefore, a heavier bounce vehicle must also maintain a similar impulse time duration. Such a vehicle would have to add more energy absorption content ahead of the *front-body-hinge-pillar* and increase the strength of the back-up structure, since the applied loads would be higher, as illustrated in Fig. 4.81b.

SOF for a BOF

A Body-on-Frame test vehicle has the added complexity of being two linked structures, as represented in side-view in Fig. 4.82. The chassis frame and body each have a strength profile ($s_f(\delta_f)$ and $s_b(\delta_b)$, respectively) and are connected to each other by bushings with a stiffness K_{Sbm}.

SOF performance for a typical BOF ICE automobile is more difficult than a BFI ICE auto because;

- a BOF vehicle tends to be larger and heavier,
- the primary front crash structure of the chassis-frame-front-rails tend to be further inboard and thus further away from the SOF barrier,

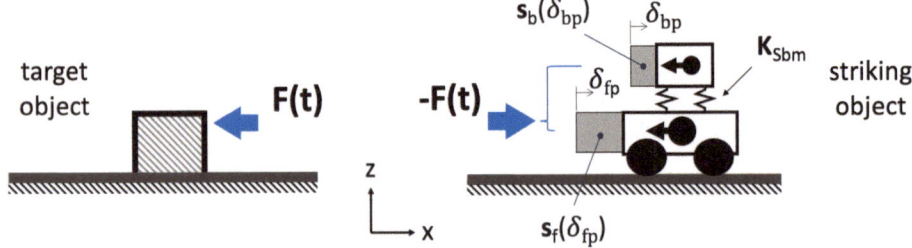

Fig. 4.82 BOF collision model

- engineering a direct loadpath to the powertrain (Fig. 4.59k) tends to be much more difficult because of the powertrain size, orientation, and packaging,
- and the front suspension geometry is usually not ideal with respect to wheel/tire kinematics and rotation.

The presence of the body bushings does provide the BOF configuration with an advantage, however. The bushings can stretch as load is applied to the front-body-hinge-pillar by the wheel/tire or direct loading by the barrier. This compliance increases the duration of the barrier/body impulse event and can moderate the force magnitude as seen by the body.

Most BOF ICE automobiles exhibit a bounce crash mode (as shown in Link-4.18), however there is at least one example where a glance mode was engineered (an example shown in Link-4.19).

SOF design elements found on a BOF automobile follow the same principles as those found on a BFI, however some design elements found in a BFI automobile are infeasible based on peculiarities associated with the BOF configuration.

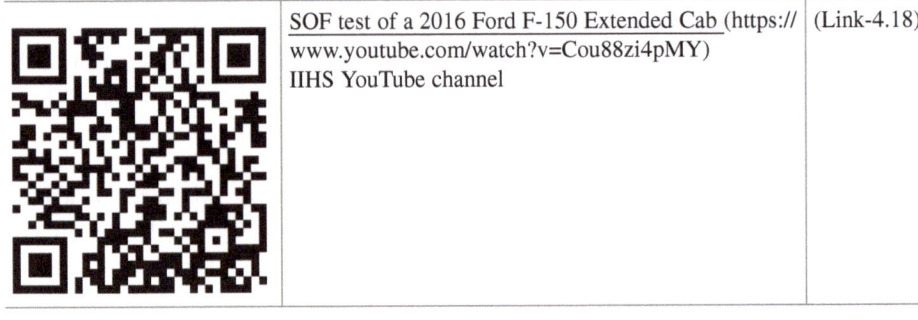

| | SOF test of a 2016 Ford F-150 Extended Cab (https://www.youtube.com/watch?v=Cou88zi4pMY) IIHS YouTube channel | (Link-4.18) |

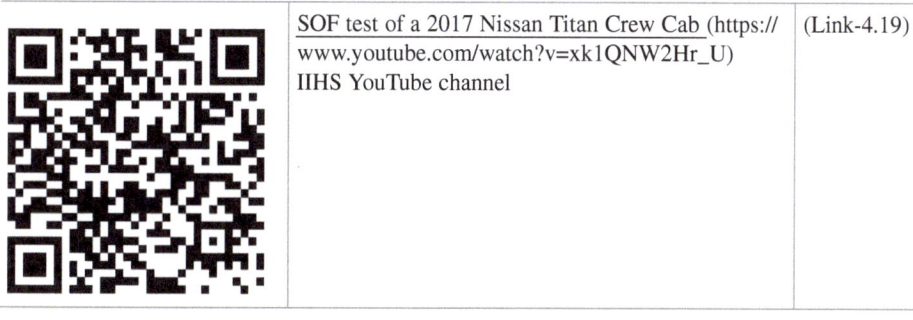

	SOF test of a 2017 Nissan Titan Crew Cab (https://www.youtube.com/watch?v=xk1QNW2Hr_U) IIHS YouTube channel	(Link-4.19)

4.3.12 Peculiarities of EVs

Mass

Since BEVs are heavier than their ICE counterparts in the era of 600 Wh/l battery cell technology, they are susceptible to the detailed conundrums described above.

In addition to a general mass increase, the presence of a propulsion battery content results in the condition where a larger percentage of the vehicle mass is concentrated behind the dash. Here, increased plan-view shear stiffness of the front compartment is required to achieve the same lateral deflection for a glance crash mode; as the mass increases, K_{Sm} must also increase to maintain performance, as illustrated in Fig. 4.83.

Post-Crash High Voltage Safety

Although FMVSS305 and UNIECE regulations do not explicitly apply to the SOF loadcase, manufacturers consider post-crash high voltage safety performance in SOF performance development for EVs.

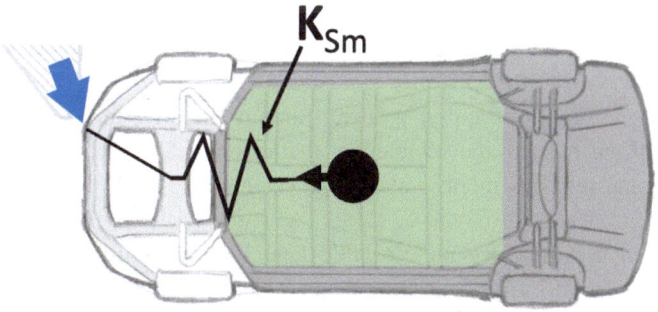

Fig. 4.83 EBVs and front-compartment shear stiffness

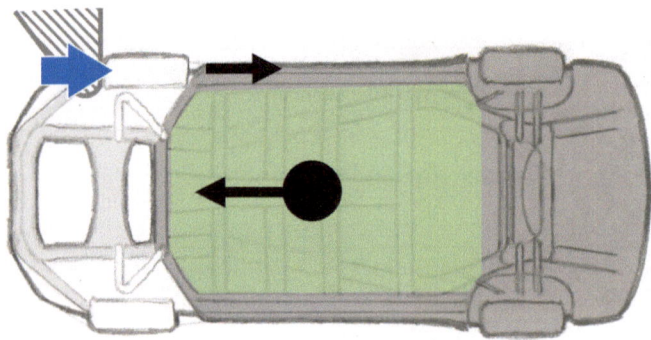

Fig. 4.84 BEVs and front-compartment shear force visualization

In particular, the near-side tire/wheel impact and intrusion is considered when packaging the propulsion battery housing. Also, the body structure and propulsion battery housing designs sometime require strengthening to manage the plan-view shearing effect caused by a large longitudinal force applied to the *front-body-hinge-pillar*, such as in a vehicle with a bounce crash mode (illustrated in Fig. 4.84).

Careful consideration of high voltage component packaging and crash strategy selection and development are hallmarks a vehicle designed efficiently for SOF.

Aspects of Electric Drive Unit

Using the electric drive unit to increase the vehicle's lateral velocity, as illustrated in Fig. 4.59k, is made difficult since the drive unit includes high voltage content and tends to be much smaller and lighter than their ICE counterparts.

- Depending on how the high voltage circuit is managed during and after a crash, manufacturers might avoid loading the drive unit simply because of its high voltage content and a motivation for post-crash high voltage safety.
- The fact that the drive units are smaller than their ICE counterparts make it more difficult to engage the drive unit with the *front-compartment-mid-rail* as they tend to be further away from the *mid-rail*.
- Ultimately, the benefit of applying lateral force to the drive-unit is diminished if it is not a significant mass in the vehicle, per the lesson of Sect. 3.2.2; a force directed at an object's masses is required to change its velocity.

The inability to use the drive unit to build lateral velocity further increases the need for plan-view shear stiffness of the front compartment, as illustrated in Fig. 4.83.

4.4 Oblique Side-Pole

Oblique Side-Pole, often simply called "Side-Pole" within the automotive design community, is a loadcase in which a test vehicle impacts a stationary, 'rigid' pole with a specified velocity and yaw angle, as illustrated in Fig. 4.85. The test fixture is designed and constructed such that friction between the test vehicle's tires and the ground during its approach is eliminated; often accomplished by having the test vehicle resting on a large plate that travels towards the pole. Link 4.20 and 4.21 provide videos of a Side-Pole test.

This loadcase can be found as a regulatory test and as consumer metric tests, however detail test parameters, such as the test vehicle's velocity, are not necessarily the same across all test protocols. The US FMVSS214 and the China GB/T37337-2019 are examples of the loadcase being a regulation while examples of the loadcase existing as an NCAP are, USNCAP, EuroNCAP, and ANCAP.

The fore-aft location of the vehicle with respect to the pole depends on the specific test protocol, as illustrated in Fig. 4.86. For example, the FMVSS214 test protocol considers two locations in two separate tests; one aligning to the seating position of a 5th percentile female occupant and another aligning to the seating position of a 50th percentile male. The ENCAP Side-Pole Test procedure, on the other hand, considers a pole contact location aligned to the vehicle's cg.

Performance in the Side-Pole loadcase is largely based on occupant performance metrics that are extracted from the crash dummies used in the test. The USNCAP rating, for example, includes metrics that predict injury to the head, spine, and ribs.

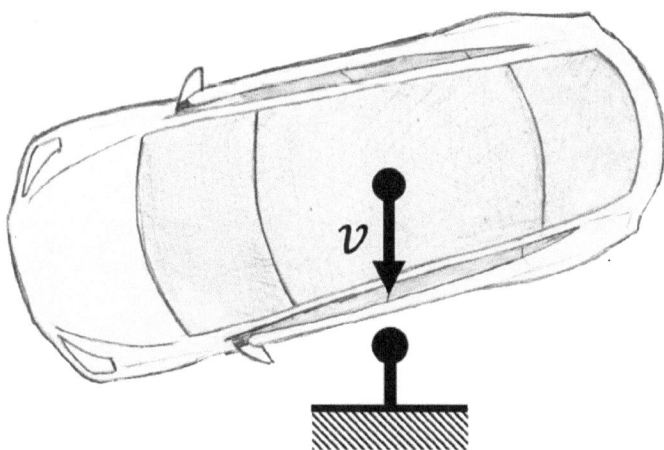

Fig. 4.85 General side-pole test condition

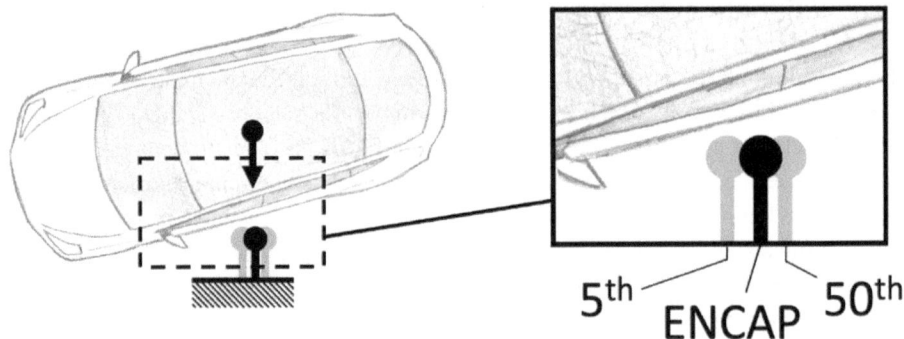

Fig. 4.86   Side-pole test locations

	Side-Pole test of a 2023 Toyota RAV4 Prime, top view NHTSA USNCAP, 32kph, 5th percentile position	(Link-4.20)
	Side-Pole test of a 2023 Toyota RAV4 Prime, front view NHTSA USNCAP, 32kph, 5th percentile position	(Link-4.21)

4.4.1   Loadcase Modeling

The test vehicle strikes a stationary 'rigid' barrier in Side-Pole, however we can utilize the collision model in the same way as in other stationary barrier loadcases by considering the target object to be the barrier and the earth that it is attached to. Here, $m_T \gg m_s$ and we would expect, and find that, the velocity change of the barrier/earth to be immeasurable.

The general collision model for Side-Pole is shown here in Fig. 4.87.

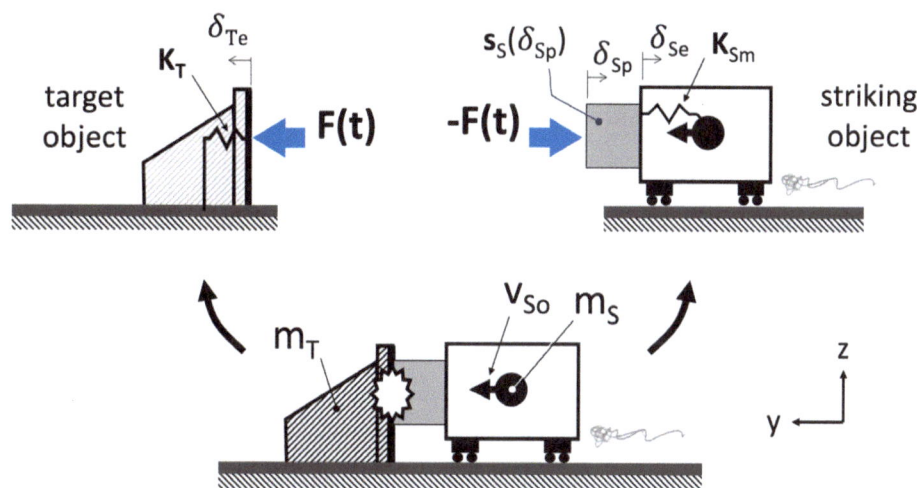

Fig. 4.87 General side-pole collision model

Fig. 4.87 variable definitions:

s_S	=	is the strength profile of the striking vehicle
δ_{Sp}	=	is the direction of striking vehicle plastic deformation measure
δ_{Se}	=	is the direction of striking vehicle elastic deflection measure,
K_{Sm}	=	is the stiffness of the striking vehicle between impulse and mass center,
δ_{Te}	=	is the direction of barrier elastic deflection measure,
K_T	=	is the stiffness of the barrier

- Plasticity (s_S): Plasticity exists only in the striking object, assuming that the barrier has been designed and constructed well.
- Friction: The presence of friction between the test vehicle and ground depends on the test fixture design, however it is a design goal of all test fixtures to minimize this friction prior to the collision event; to the point which it can be ignored. During the collision, friction occurs between the striking object and the ground in cases where the striking object rotates in plan-view as it collides with the pole. Its effect is usually ignored and the cause of plan-view rotation is discussed later in this chapter.
- Elasticity (K_T, K_{Sm}): Although the pole is referred to as "rigid", it does have elasticity (K_T), If the barrier is designed and constructed well, its deflection will be extremely small. The elasticity of the striking vehicle between the applied impulse force and its generalized mass center (K_{Sm}) is most relevant in this loadcase.

Insight from Governing Equations

The impulse equation can be leveraged from out collision model developed in Sect. 3.2.3.5. Replicated here, impulse is defined as the change in a collision object's momentum (Eq. 4.77), but also equivalent to the area under the F(t) curve (Eq. 4.78).

$$\text{Impulse}: \boxed{J = p_o - p_f} \qquad [\text{N s}] \text{ or } [\text{kg m/s}] \tag{4.77}$$

$$\text{Impulse}: \boxed{J = \int_o^f F(t)dt} \tag{4.78}$$

or

$$\boxed{J = F_{ave}\Delta t} \tag{4.79}$$

where:

J	=	impulse
p	=	momentum of a collision object
F	=	impulse force
Δt	=	time duration of collision

...equating these two...

$$J = J \tag{4.80}$$

$$F_{ave}\Delta t = p_o - p_f \tag{4.81}$$

...executing the same steps performed for the Full-Frontal loadcase model

$$\boxed{F_{ave}\Delta t = m_S v_{So}} \tag{4.82}$$

where:

m_S	=	mass of the striking object
v_{So}	=	initial velocity of the striking object
F_{ave}	=	average impulse force
Δt	=	time duration of collision

Equation 4.82 shows us that the F·t product required to slow the striking object is dependent on its mass. This should feel intuitive, as a heavier vehicle will have more initial kinetic energy and slowing it to a stop will require more effort.

We can also employ the conservation of energy law…

$$E_f = E_o \tag{4.83}$$

$$\left(KE_{Sf} + W \right) = KE_{So} \tag{4.84}$$

$$\boxed{KE_{Sf} + F_{ave}\Delta_{Sp} = \left(\frac{1}{2}m_S v_{So}^2 \right)} \tag{4.85}$$

where:

m_S	=	mass of the striking vehicle
v_{So}	=	initial velocity of the striking vehicle
KE_{Sf}	=	residual kinetic energy of the striking vehicle
F_{ave}	=	average impulse force
Δ_{Sp}	=	crush distance seen within striking vehicle

We find in Eq. 4.85 that some of the striking object's initial kinetic energy will be converted into material fracture and crush within the test vehicle. The amount of crush is dependent on the force capacity of the test vehicle; as the force capacity increases, the average impulse force increases and the amount of crush decreases.

One might think that KE_f is zero, however this is only true when the striking object's cg and the pole are perfectly aligned. When there is a fore-aft offset between the cg and the pole, KE_f exists as the striking object's rotational velocity (Fig. 4.94).

System Energy Flow

Most of the system energy will be converted into crush in the striking vehicle, as illustrated in Fig. 4.88. The striking vehicle might have residual kinetic energy in the form of rotational velocity, depending on how its cg aligns with the pole. Some energy is stored during the event. The amount of stored energy is dependent on the stiffness characteristics of the pole and the striking vehicle (K_T and K_{Sm}, respectively) and the maximum elastic compression seen in the two objects (Δ_{Te} and Δ_{Se}, respectively). The stored energy is released once the impulse force is removed and results in the striking vehicle 'springing' off the pole.

4.4.2 Fundamental Behavior and Physics

The FBD of the test vehicle is shown in Fig. 4.89 and includes the elements of its initial momentum and the impulse force which will change its momentum during the collision.

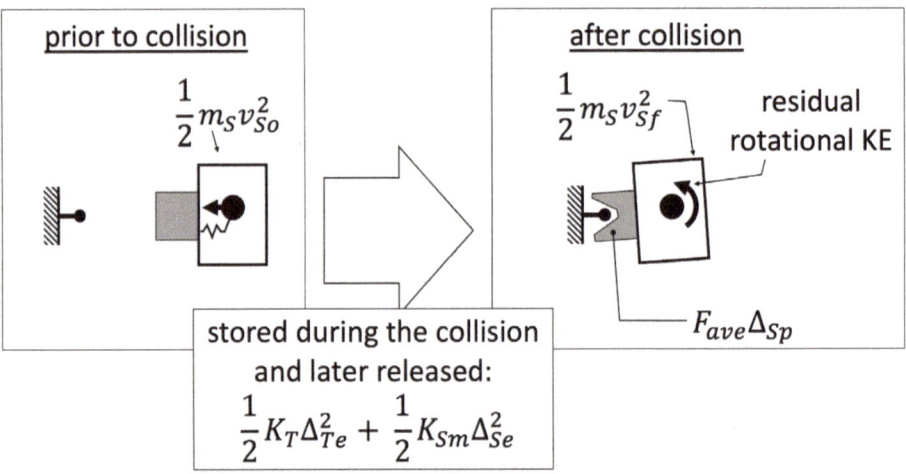

Fig. 4.88 Side-pole energy conservation

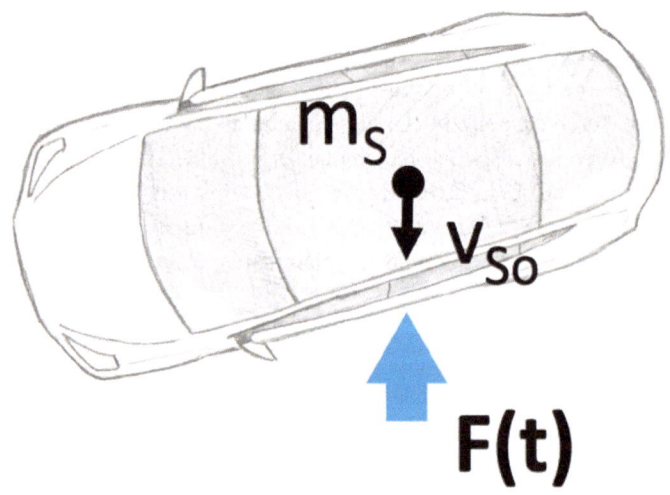

Fig. 4.89 Side-pole test vehicle FBD

Event Milestones

Typical vehicle behavior in the Side-Pole loadcase is illustration in Fig. 4.90.

After initial contact with the barrier, crush in the test vehicle begins. The structure begins to store elastic energy as the impulse force is applied. The vehicle might exhibit a yaw rotation based on a longitudinal misalignment between the vehicle cg and the pole location (Fig. 4.92). The vehicle might also exhibit a roll rotation based on a vertical misalignment between the vehicle cg and the rocker location (Fig. 4.96). The vehicle yaw

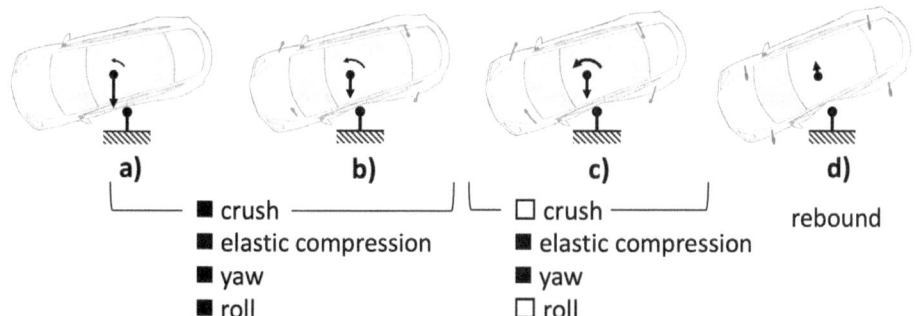

Fig. 4.90 Side-pole event milestones

will be resisted by friction between the tire and ground. At the end of the collision event, the stored elastic energy is released and 'springs' the vehicle off the barrier.

These aspects of the collision event are particularly important with regards to understanding structural behavior: energy absorption through crush, yaw, ride-down of the test vehicle mass, roll, and spring-back. The complexity of behavior in Side-Pole makes it appropriate that discussion focus on these aspects individually, rather than focusing on the stages of the event.

Crush

Material fracture and crush within the vehicle and its structure will occur as the impulse force overwhelms individual component material properties. Crush starts in the side-door system, as this is the first part of the vehicle to contact the pole. Crush will extend to the body side structure once it contacts the pole.

The acceptable amount of crush and the acceptable time duration of the collision event is complicated, as performance in the Side-Pole loadcase is largely determined by occupant performance measures and occupant performance is dependent both on the vehicle crush magnitude and deceleration and the occupant's collision with the vehicle itself (the 'other collision', described in Sect. 3.3).

When developing structure in the context of Side-Pole occupant performance, it is important to remember the relationships learned from developing the collision model and one from general physics:

- as the collision time duration reduces, the force increases (Eq. 4.82)
- as the crush distance reduces, the force increases (Eq. 4.85)
- as the collision time duration reduces, the deceleration increases ($a = v(t)\, dt$)

Fig. 4.91 2016 Volvo S60 post-test, USNCAP side-pole

For a typical crush magnitude, consider the 2016 Volvo S60 tested under the USNCAP Side-Pole protocol (32kph, 5th percentile female pole location, 1848 kg test weight); shown in Fig. 4.91. The pole intrusion values measure between 300 and 400 mm at different vertical locations on the door. The force and acceleration applied to the S60's occupant was low enough that this Volvo achieved the maximum score of '5-star' for occupant protection.

Yaw

Note that, in Fig. 4.89, the mass of the test vehicle is represented at the vehicle's cg and is shown in line with the pole. Because the impulse force and the deceleration force are aligned, no yaw rotation of the vehicle would occur in this condition. Figure 4.92a illustrates the condition where there is a fore-aft offset between the vehicle's cg and the pole; yaw rotation would certainly be expected in such conditions.

The test vehicle will have some residual kinetic energy in the form of test vehicle rotational velocity when the cg does not align to the pole. The condition where the vehicle cg does align with the pole location is therefore the most difficult test condition for the vehicle's structure to manage, as all the initial kinetic energy is converted to material fracture and crush.

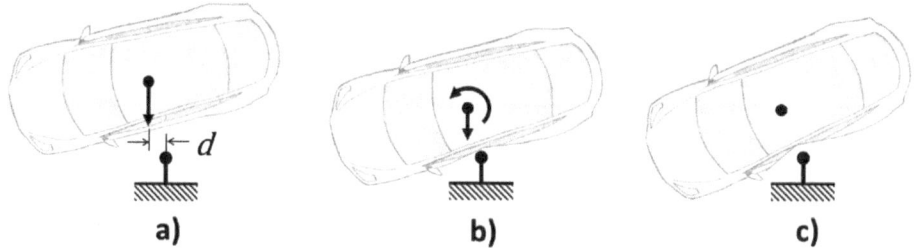

Fig. 4.92 Mechanism of vehicle rotation

Energy Ride-down

Recall from Eq. 4.77 that impulse is the process of changing a collision object's momentum, and particularly changing the velocity of a collision object's mass. Considering that, it is appropriate to explore the FDB at a more granular level, recognizing that there are several significant masses within the vehicle that need to be slowed during the collision and impulse event. Consider the detailed FBD shown in Fig. 4.93 where the mass and velocity of some heavily components are represented.

During the Side-Pole collision event, forces are applied to the structure as these components decelerate, as illustrated in Fig. 4.94. This uncovers the importance of recognizing the vehicle structure between heavy components and the pole; buckling and deformation can occur within the structure due to the internal loads applied by heavy components and can have implications on intrusion at the occupant location.

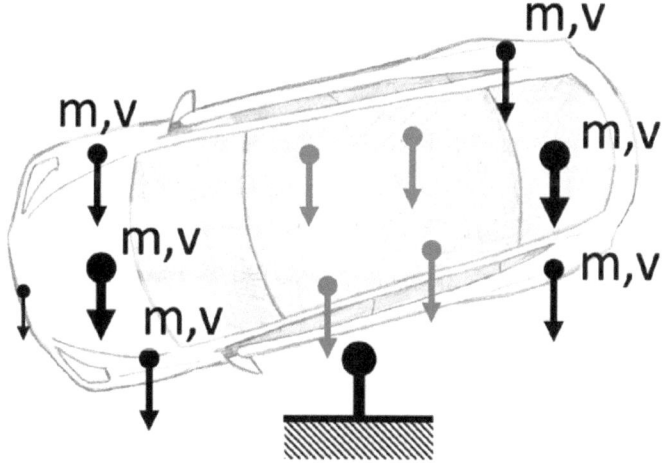

Fig. 4.93 Detailed FBD of side-pole

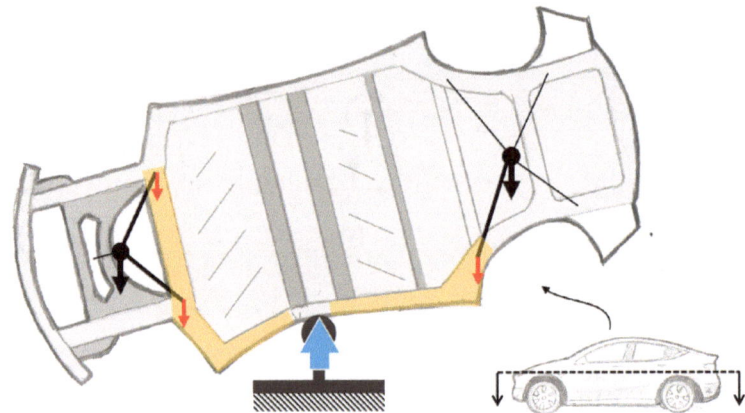

Fig. 4.94 Ride-down loadpath within the structure

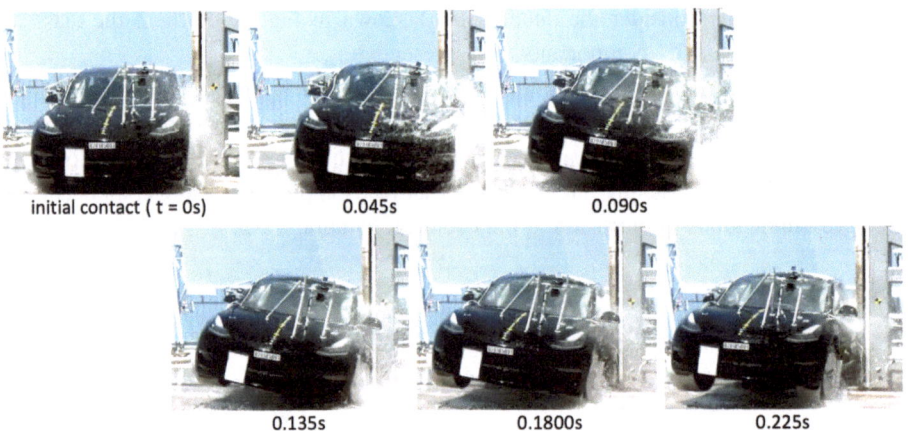

Fig. 4.95 Vehicle role during side-pole testing

Roll

Video of Side-Pole crash tests nearly always show the vehicle rolling as it strikes the pole barrier. The image sequence of the 2018 Tesla Model-3 USNCAP test shown in Fig. 4.95 illustrates this behavior. The behavior can also be seen in the video referenced by Link 4.22.

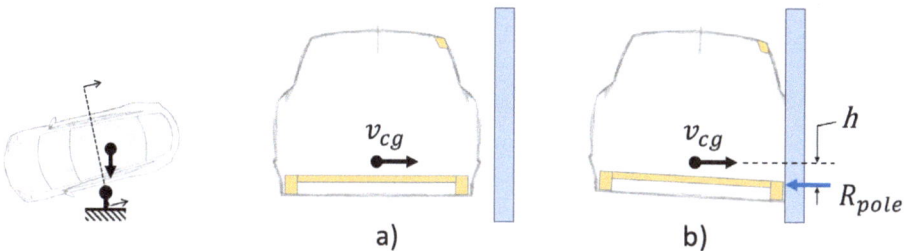

Fig. 4.96 Mechanism of vehicle roll

	Side-Pole test of a 2018 Telsa Model-3, front view	(Link-4.22)
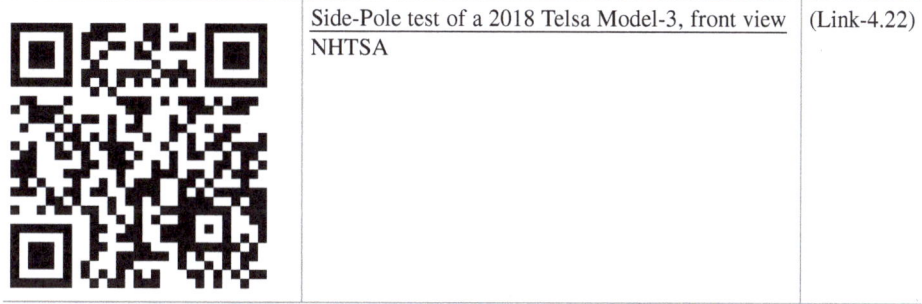	NHTSA	

Examining the FBD in front view (Fig. 4.96) explains this behavior, as we see there is a vertical offset between the vehicle's center of gravity and the vehicle's rocker. Although the side-door contacts the pole prior to the rocker, it is the rocker that has a stiff and strong loadpath to the vehicle mass elements and can therefore instigate a meaningful impulse force (recall the lessons and experiments of Sects. 3.2.3.2 and 3.2.3.3). This rotation will occur until the impulse force is relatively low or the vehicle rolls to the point where a structural element above the cg, contacts the barrier; this could be the *roof-rail*.

Rebound
Like in other crashworthiness loadcases, the force applied to the test vehicle results in a level of elastic compression of the striking vehicle and barrier, as detailed in Sect. 3.2.3.2. The stored energy is released as the impulse force reduces and causes the test vehicle to 'spring off' of the barrier at the very end of the event.

4.4.3 Structural Loadpath Topology

The loadpath topology for Side-Pole typically consists of the front side-door structure, the body structure between the pole and occupant, and the body structure between the pole and heavy components within the vehicle, as illustrated by Fig. 4.97.

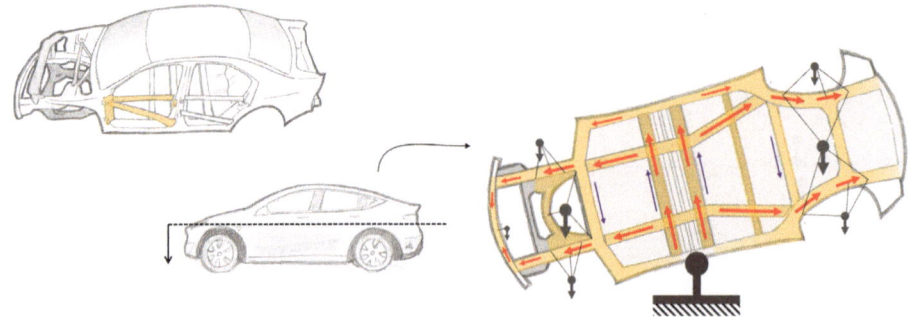

Fig. 4.97 Typical side-pole structural loadpath topology

Since the crush distance tends to be small, even the elements that see plastic deformation during the event, such as the side-door structure, have a significant load distribution function.

4.4.4 Peculiarities of EVs, Side-Pole

Mass
BEVs in the era of 600 Wh/l can be more than 500 kg heavier than their ICE counterpart and this increased mass affects the impulse magnitude required to slow the test vehicle. Per the impulse equation from our collision model (Eq. 4.82, duplicated below as 4.86), the F·t product will need to compensate.

$$F_{ave}\Delta t = m_S v_{So} \tag{4.86}$$

Ultimately, the structure will be required to manage the higher impulse forces during the collision.

Post-Crash High Voltage Safety
Although FMVSS305 and UNIECE regulations does not explicitly apply to the Side-Pole loadcase, post-crash high voltage safety performance is considered in the design and testing of EVs.

A BEV automobile with a full-width underfloor propulsion battery configuration often have battery cells and other high voltage components positioned further outboard than the occupant. These components can be sensitive to crush, depending on their chemistry and design. Figure 4.98 illustrates that the outboard position of these components can require a reduction in allowable intrusion relative to an ICE automobile.

As we know from the Conservation of Energy, a reduction of allowable intrusion (Δ_{Sp}) equates to a higher force.

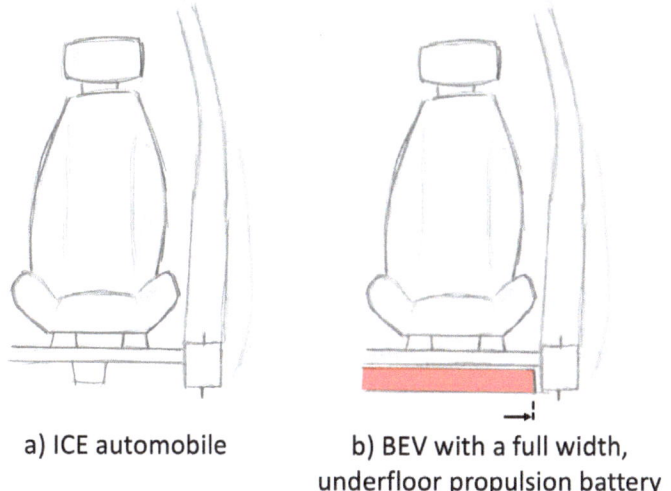

a) ICE automobile   b) BEV with a full width,
                                     underfloor propulsion battery

Fig. 4.98   ICE versus BEV in section

$$F_{ave}\Delta_{Sp} = KE_{So} - KE_{Sf} \tag{4.87}$$

Elimination of Body Ride-Down Loadpaths

Although some BEVs with a full-width underfloor propulsion battery configuration can package body center-compartment-mid-rails above the floor (BEVs with a high seating position, in particular), packaging the underfloor battery necessitates removal of center-compartment-mid-rails in most designs, as illustrated in Fig. 4.99. The loss of these loadpaths not only complicates the structure's ability to manage the higher collision forces associated with a heavier vehicle, they make it more difficult to manage the ride-down loads generated by components. Specifically, elimination of these structural elements diminishes the capability to manage the shear loading that occurs, as represented by purple arrows in Fig. 4.99.

Implications to Topology-Ride Down Loadpaths

Elimination of the body *center-compartment-mid-rails* creates a condition where the body structure is reliant on a ladder frame configuration, however the situation is further complicated by the buckling behavior of the near-side rocker. With the understanding that collapsed or bucked structural elements are poor at transmitting bending loads, the body structure loadpath topology for managing ride-down of components is essentially that illustrated in Fig. 4.100; exceptionally susceptible to shear deformation as represented by purple arrows.

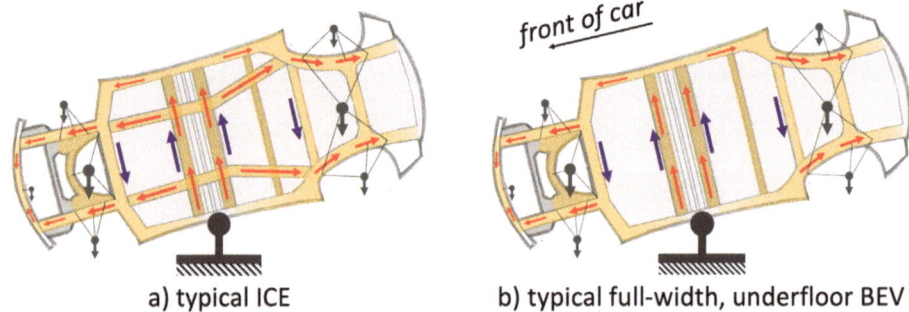

Fig. 4.99 Center compartment implications

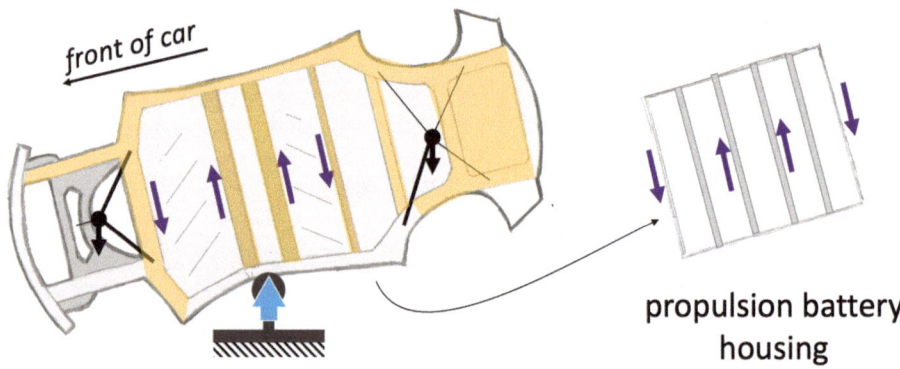

Fig. 4.100 Shear load condition within the structure

The challenge for full-width underfloor BEVs is to maximize the body structure's ability to manage shear and to develop propulsion battery housing designs that supplement the body structure to boost plan-view shear strength. Careful consideration of the structural topology during the early stages of the design development process is required.

Some BEVs are notable for their innovation with regards to designs to counteract the loss of the center-compartment-mid-rails. The 2018 Tesla Model-3 oriented the internal propulsion battery housing structure longitudinally such that it essentially relocated the *center-compartment-mid-rail* from the body to the propulsion battery housing, as illustrated in Fig. 4.101a. The 2022 GMC Hummer EV maximized the body structure floor panel's ability to transmit shear by (nearly) eliminating any geometric features that make the panel non-planar. The 2022 Tesla Model-Y incorporated a design that leveraged the propulsion battery cells as a shear loadpath by bonding them together and carefully integrating that loadpath into the body structure, as illustrated in Fig. 4.101b.

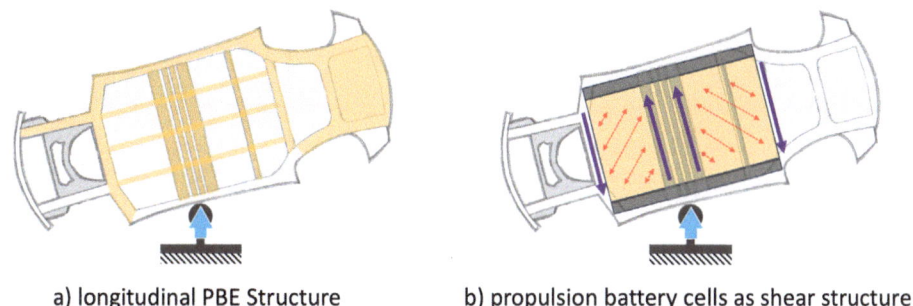

a) longitudinal PBE Structure   b) propulsion battery cells as shear structure

Fig. 4.101   Innovative solutions to manage shear loading

Implications to Topology, Side Structure

Engineering successful structural designs becomes challenging as the allowable crush space reduces to the point at which it is essentially the width of the vehicle's rocker (as illustrated in Fig. 4.102). Helpful insight can be from the Conservation of Energy law:

$$KE_o = KE_f + W \qquad\qquad (4.88)$$

...assuming that residual kinetic energy in the form of rotational velocity is negligible and recognizing that by using "≈"...

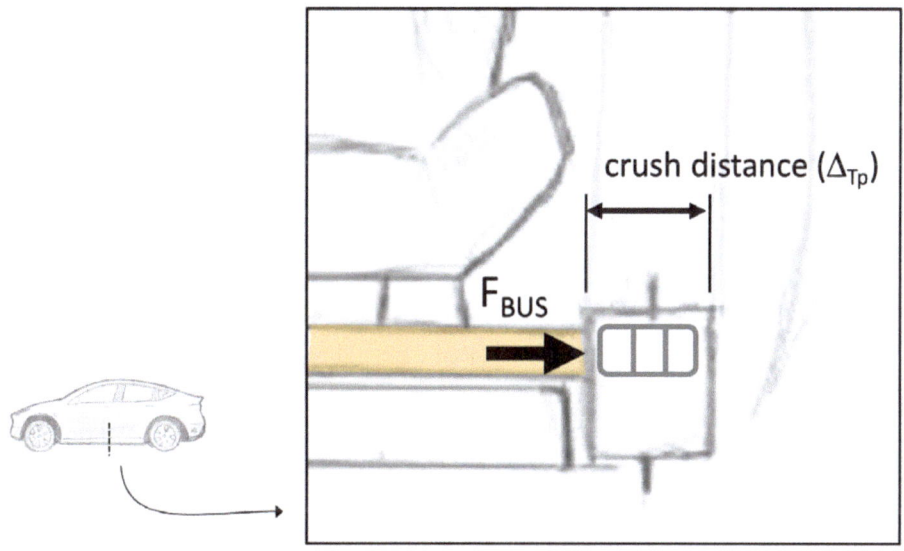

Fig. 4.102   BEV section forward of B-pilar

$$KE_o \approx 0 + W \tag{4.89}$$

$$\frac{1}{2}m_S v_S^2 = F_{BUS} \times \Delta_{Tp} \tag{4.90}$$

$$\boxed{F_{BUS} = \frac{m_S v_{So}^2}{2\Delta_{Tp}}} \tag{4.91}$$

where:

m_S	=	mass of the striking vehicle
F_{BUS}	=	force applied to back-up structure
m_S	=	mass of the striking vehicle
F_{ave}	=	initial velocity of the striking vehicle
Δ_{Tp}	=	crush distance seen within target vehicle

Equation 4.91 uncovers a relationship between the crush distance and the force applied to the back-up structure. This is an important relationship to understand during the vehicle design development process; the force that the structure is required to manage increases as the allowable crush space decreases. And, at some reduced level of allowable crush space, the force is so high that a structural solution is infeasible.

To put this into context, consider the hypothetical example shown in Table 4.2. Here, simulation and calculation indicate that 140 mm of crush distance is required to have a reasonable force level for the back-up structure to manage; higher loads would require more structural content than would otherwise be required, considering all the vehicle loadcases.

Engineering the vehicle to a crush distance of 90 mm would increase the force on the back-up structure by 56%, driving more mass and cost into the back-up structure's design solution. Finding the right solution for a vehicle often requires balancing battery volume with structural content.

Table 4.2 Effect of allocated crush distance

m_S = 2200 kg	
v_{So} = 8.9m/s (32 kph) (20 mph)	
Baseline	*Proposal*
Crush space (Δ_{Sp}) = 140mm	Crush space (Δ_{Sp}) = 90mm
$F_{BUS} = \frac{m_S v_{So}^2}{2\Delta_{Tp}}$	$F_{BUS} = \frac{m_S v_{So}^2}{2\Delta_{Tp}}$
$F_{BUS} = \frac{(2200)(8.9)^2}{2(140)}$	$F_{BUS} = \frac{(2200)(8.9)^2}{2(90)}$
$F_{BUS} = 622KN$	$F_{BUS} = 968KN$

Implications to Topology, Back-up Structure Positioning

As we saw earlier in this section, vehicles tend to roll as the rocker contacts the barrier due to the position of the cg. This behavior encourages the rocker to rotate with respect to the vehicle and, if the structural topology does not sufficiently resist the rocker rotation, the rocker will tend to encroach in the upper corner of the propulsion battery assembly (A in Fig. 4.103).

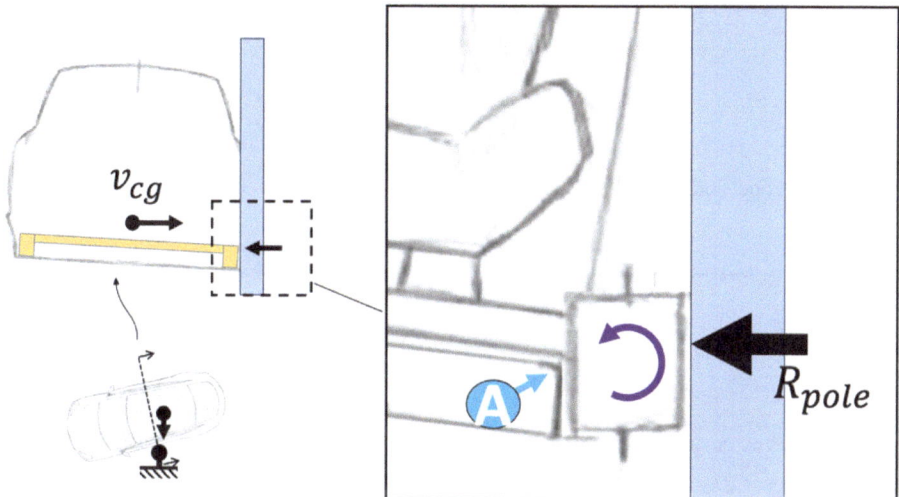

Fig. 4.103 Rocker roll and propulsion battery housing intrusion

Most, if not all, early BEVs employed a ladder frame construction within the propulsion battery housing, as illustrated in Fig. 4.104 and it is tempting to leverage the PBH-bars as back-up structure for the outboard crush elements with this configuration, as illustrated in Fig. 4.105a.

Due to the roll tendency, however, it is more efficient to support the rocker as high as possible, minimizing the vertical offset between the cg and the rocker. Figure 4.105b illustrates an example of this, where a strong lateral member is positioned to support the top of the rocker.

Implications to Occupant Performance, "the Other Collision"

As we have seen, a BEV with less the allowable crush distance relative to their ICE counterpart must manage a higher back-up structure force, per Eq. 4.91. Equation 4.82 tells us that the higher back-up structure force results in a shorter impulse duration; A BEV with less crush space than its ICE counterpart will slows down quicker than that ICE. This higher deceleration can make the job of engineering occupant performance more difficult. Recall the lessons of the 'jumping into bed' example in Sect. 3.3; slowing down quicker increases the applied force.

Fig. 4.104 Propulsion battery housing with a ladder-frame topology

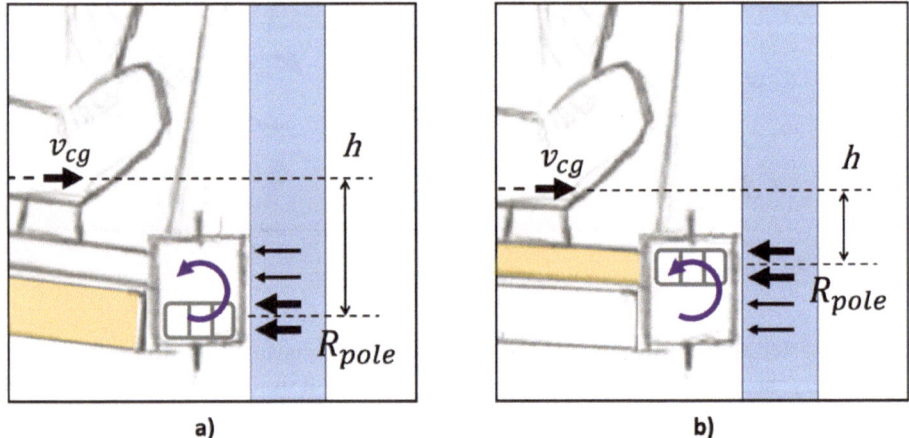

Fig. 4.105 Maximizing loadpath efficiency

Equating reduced intrusion to better occupant performance can sometimes be intuitive, as the "Under the Façade" story describes, but it is not necessarily the case for dynamic events.

Under the Façade—The Relationship Between Collision Time Duration and Acceleration

During a 2017 Tesla event to celebrate the first customer deliveries of the Tesla Model-3, Elon Musk touted the safety performance of the automobile by comparing the Model-3 Side-Pole test video to that of a 2016 Volvo S60; the S60 being a recognizable safety benchmark, having scored a 5-star rating in the USNCAP Side-Pole test and receiving a 5-star rating for its USNCAP overall score

Musk's comparison showed that the Model-3 had significantly less pole intrusion than the S60 and crowd went wild, assuming that occupant performance was a function of intrusion

He did not explain that the Model-3's intrusion is less than the ICE Volvo because the propulsion battery needs to be protected and let the crowd carry-on with their intrusion-based safety conclusion, adding, "It is obvious which car you would prefer to be in, in an accident"

Musk's statement is not necessarily false if he had been referring to the fine details of the other USNCAP loadcases, however the injury metrics for the Side-Pole test that NHTSA performed exemplify that 'stopping quicker is not necessarily a good thing'. Comparing the Side-Pole loadcase injury metrics between these two vehicles show that the Volvo outperformed the Model-3 in four out of the five injury metrics

Measure	Threshold	2018 Tesla Model-3 NHTSA test 10385	2016 Volvo S60 NHTSA test 9496
Head injury criteria (HIC$_{36}$)	1000	384	**251**
Resultant lower spine Acceleration	82	41 g	**33 g**
Total pelvic force (acetabular + iliac)	5525	**2478 N**	3140 N
Maximum thoracic rib Deflection	38	23 mm	**19 mm**
Maximum abdomen rib Deflection	45	28 mm	**22 mm**

4.5   Side Moving Deformable Barrier

There are several Side Moving Deformable Barrier (MDB) loadcases around the world, both regulatory and consumer metric tests. All tests involve a stationary test vehicle being struck by a deformable barrier attached to a striking vehicle (Fig. 4.106). There are many variations of the test condition, however. Differences in test protocols include; the striking vehicle speed, the striking vehicle weight, the height of the barrier, the vector in which the striking vehicle travels, and even the attributes of the barrier face itself. The video referenced by Link 4.23 provides an example of the IIHS Side-MDB test condition.

The protocol for these tests attempts to replicate the worst case scenario with respect to the longitudinal location of impact on the struck vehicle. Here, the barrier is positioned such that it strikes between the *front-body-hinge-pillar* and *lock-pillar*; minimizing participation of the body structure.

Details of the rating system for each test varies but occupant performance, as measured by acceleration readings from the dummy, and structural intrusion measures are generally considered.

Fig. 4.106   General side MBD test condition

	Side-MDB test of a 2022 Honda Accord (https://www.youtube.com/watch?v=L7Vw0dAvqnE) IIHS YouTube channel, protocol 2	(Link-4.23)
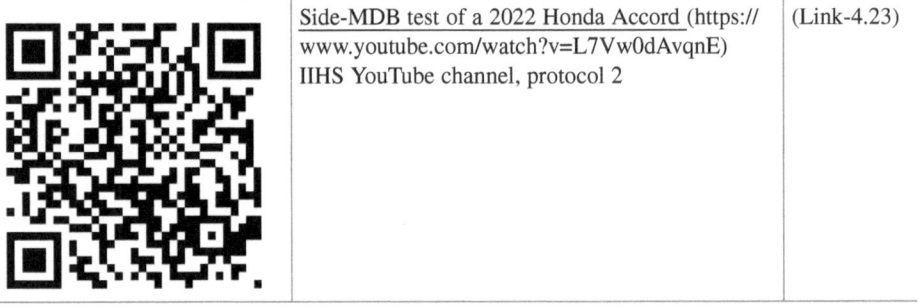		

4.5.1   Loadcase Modeling

The stationary test vehicle is struck in the side by a barrier moving at a specified velocity. The general collision model for Side MDB is shown here in Fig. 4.107.

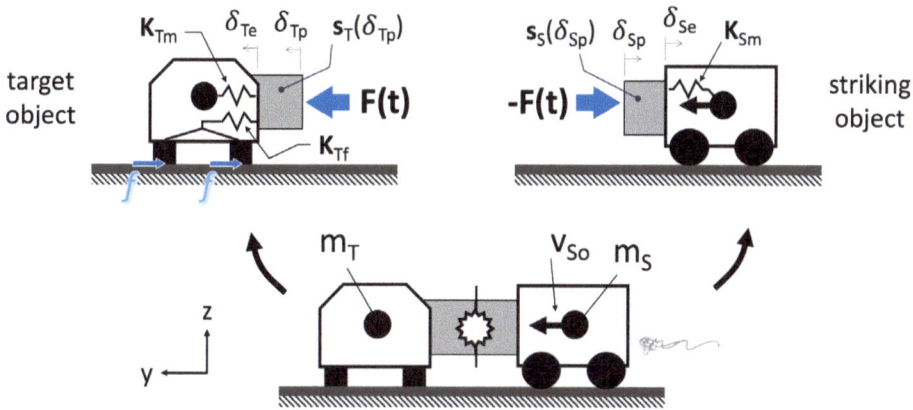

Fig. 4.107 General Side-MDB collision model

Fig. 4.107 variable definitions:

s_S	=	is the strength profile of the striking vehicle,
δ_{Sp}	=	is the direction of striking vehicle plastic deformation measure,
δ_{Se}	=	is the direction of striking vehicle elastic deflection measure,
K_{Sm}	=	is the stiffness of the striking vehicle between impulse and mass center,
s_T	=	is the strength profile of the target vehicle,
δ_{Tp}	=	is the direction of target vehicle plastic deformation measure,
δ_{Te}	=	is the direction of target vehicle elastic deflection measure,
K_{Tm}	=	is the stiffness of the target vehicle between impulse and mass center,
K_{Tf}	=	is the stiffness of the target vehicle between impulse and tire patch

- Plasticity (s_T, s_S): Plasticity exists in both the target and striking objects. The distribution of plastic deformation will depend on the relative strength of the two objects, $s_T(\delta_T)$ versus $s_S(\delta_S)$.
- Friction (f): Friction is present between the test vehicle and ground, as the tires will slide rather than roll during the collision event.
- Elasticity (K_{Tm}, K_{Tf}, K_{Sm}): Elasticity exists in both the target and striking objects (K_{Tm} and K_{Sm}, respectively). The stiffness between the applied impulse force and the vehicle's generalized mass center that is relevant for both vehicles. In this loadcase, the target vehicle stiffness between the applied impulse force and the tire patch is also relevant (K_{Tf}).

Insight from Governing Equations

Recall that no external forces act on the colliding objects in a true collision, so the presence of friction between the target object and the ground will force us to approach our equation modeling differently. First, we must subdivide the event into collision and post-collision time durations, as illustrated by Fig. 4.108. The collision event, where crush occurs, occurs between the 'initial' time and the 'final' time. Residual KE exists after a Side-MDB collision and this energy is converted into work via friction until both objects come to a complete stop at the 'end' time.

The presence of friction will also require us to append our equation modeling.

Starting with an examination of the FBD illustrated in Fig. 4.109, we find that the event impulse must account for the target object's momentum change and also resistivity of the frictional force.

We will therefore start our momentum modeling equation with the definition of impulse

$$\text{Impulse}: \quad \boxed{J = p_o - p_f} \qquad [\text{N s}] \text{ or } [\text{kg m/s}] \qquad (4.92)$$

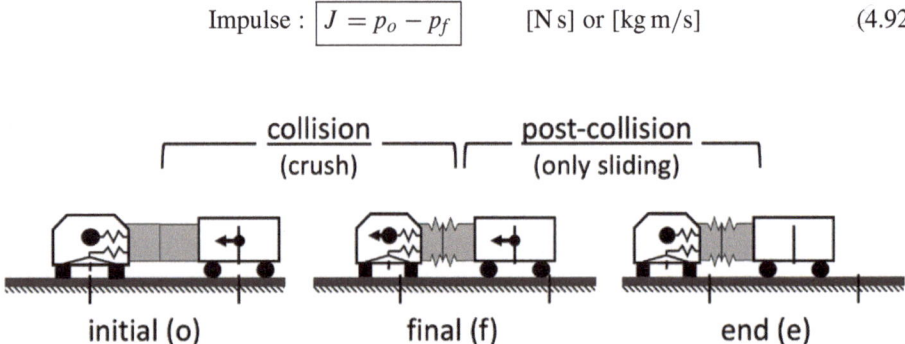

Fig. 4.108 Event sub durations and corresponding subscripts

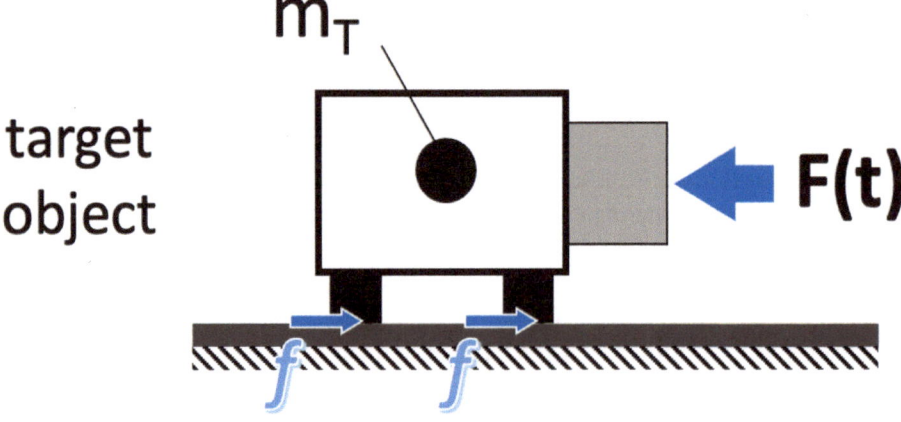

Fig. 4.109 Side-MDB FBD

$$\text{Impulse}: \boxed{J = \int\limits_{o}^{f} F(t)dt} \tag{4.93}$$

or

$$\boxed{J = F_{ave}\Delta t} \tag{4.94}$$

where:

J	=	impulse
p	=	momentum of a collision object
F_{ave}	=	collision impulse force
Δt	=	time duration of collision

An impulse equation can be written to account for the fact that the collision impulse force (J_c) must change the momentum of the target object (Δp_T) and overcome the friction (J_f); shown here in Eq. 4.95...

$$J_c = \Delta p_T + J_f \tag{4.95}$$

$$\text{where } J_f = \int\limits_{o}^{f} f(t)dt \text{ or } J_f = f_{ave}\,\Delta t \tag{4.96}$$

...substituting 4.93 and 4.96 into Eq. 4.95 we find...

$$F_{c\_ave}\,\Delta t_c = \left(m_T v_{Tf} - m_T v_{To}\right) + f_{ave}\,\Delta t_c \tag{4.97}$$

...and recognizing that the target vehicle is stationary at the start of the test...

$$\boxed{F_{c\_ave}\,\Delta t_c = m_T v_{Tf} + f_{ave}\,\Delta t_c} \tag{4.98}$$

where:

F_{c_ave}	=	average impulse force of the collision
f_{ave}	=	average frictional force
Δt_c	=	time duration of collision
m_T	=	mass of the target vehicle
v_{Tf}	=	velocity of the target vehicle just after the collision

Equation 4.98 tells us a couple things:

1. As the mass of the target vehicle increases, the F·t product of the collision will increase.
2. As the frictional force increases, the F·t product of the collision will also increase.

We can also look at this event from the perspective of the conservation of energy law...

$$E_f = E_o \tag{4.99}$$

$$KE_{Tf} + KE_{Sf} + W_T + W_S = KE_{To} + KE_{So} \tag{4.100}$$

$$KE_{Tf} + KE_{Sf} + W_T + W_S = KE_{So} \tag{4.101}$$

$$W_T + W_S = KE_{So} - KE_{Tf} - KE_{Sf} \tag{4.102}$$

...substituting and using "≈" to recognize that energy loss due to friction is not accounted for and recognizing that the two vehicles will have the same velocity at the end of the collision...

$$\boxed{F_{c\_ave} \Delta_{Tp} + F_{c\_ave} \Delta_{Sp} \approx \frac{1}{2} m_S v_{So}^2 - \frac{1}{2}(m_T + m_S) v_f^2} \tag{4.103}$$

where:

F_{c_ave}	=	average collision impulse force
Δ_{Tp}	=	crush distance seen within target vehicle
Δ_{Sp}	=	crush distance seen within striking vehicle
m_S	=	mass of the striking vehicle
m_T	=	mass of the target vehicle
v_{So}	=	initial velocity of the striking vehicle
v_f	=	velocity of the two vehicles immediately after the collision

...we find in Eq. 4.103 that some of the initial kinetic energy of the striking object is converted into crush in the barrier on the target object and crush in the striking object. The amount of crush is dependent on the force capacity of the barrier and test vehicle; as the crush zone force capacities increase, the average impulse force increases and the amount of crush decreases.

Equation 4.103 also indicates that the force·intrusion magnitude will increase as the mass of the target vehicle increase.

System Energy Flow
Figure 4.110 illustrates paths of system energy flow during the collision event;

- Some of the initial kinetic energy of the striking vehicle is converted into crush; in the target vehicle ($F_{c\_ave}\Delta_{Tp}$) and the striking vehicle ($F_{c\_ave}\Delta_{Sp}$).
- Some energy is stored during the event and released after the impulse force is removed. The amount of stored energy is dependent on the stiffness of the target and striking vehicles (K_{Tm} and K_{Sm}, respectively) and the maximum elastic compression seen within the two vehicles (Δ_T and Δ_S).
- There is residual kinetic energy associated with the collision and will be exhausted through the action of post-collision sliding friction.

4.5.2 Fundamental Behavior and Physics

Figure 4.111 illustrates the FBD of the target vehicle with representation of elasticity and plasticity. The collision force is resisted by the target vehicle mass and the frictional force that occurs between the target vehicle's tires and the ground.

Event Milestones
The typical, general behavior of the target and striker vehicles is illustrated in Fig. 4.112 and described below.

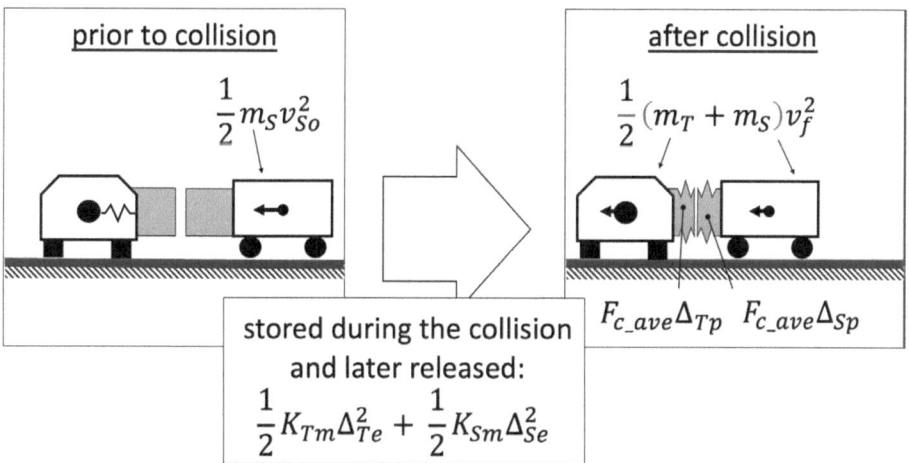

Fig. 4.110 Side-MDB energy conservation

Behavior, High-Level
Stage-1: Stationary Crush

  (a) The striking vehicle first contacts the test vehicle.
(a-b) The barrier portion of the striking vehicle and portions of the target vehicle crush while the target vehicle remains stationary. The velocity of the striking vehicle reduces as the crush dissipates energy.
  (b) The force seen by the test vehicle's mass is sufficient to begin accelerating the test vehicle and the force seen at the tire patch meets or exceeds the resistant frictional force.

Stage-2: Crush with Target Vehicle Velocity

(b-c) The test vehicle begins to move, however a frictional force resist the vehicle's motion. The velocity of the striking vehicle decreases and the velocity of the target vehicle increases.
  (c) Crush ends when the two vehicles have the same velocity; the collision event is over at this point. Release of the stored elastic energy occurs as the impulse force reduces.

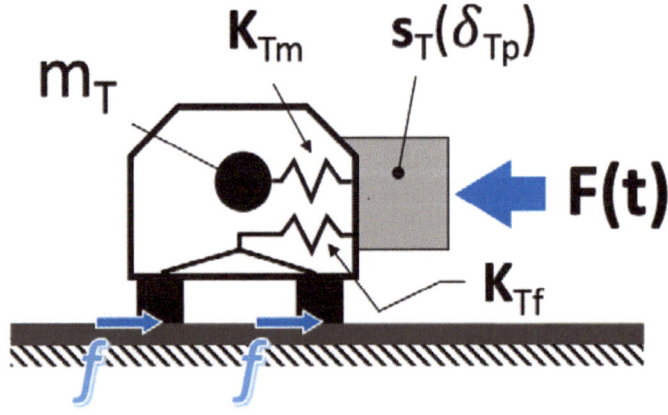

Fig. 4.111 Side-MDB target vehicle FBD

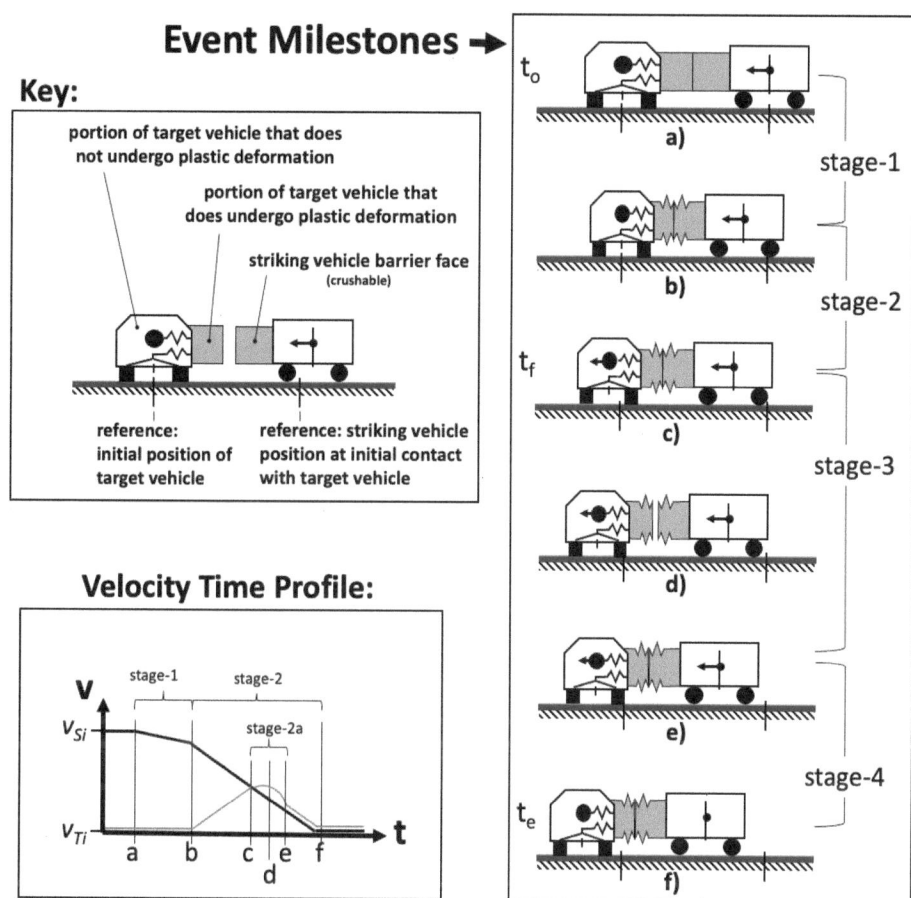

Fig. 4.112 Side MDB event milestones

Stage-3: Post-Collision, Rebound and a Second Collision (Optional)

(d) Energy has been stored in the two vehicles during the event and is released as the impulse force is removed, as detailed in Sect. 3.2.3.2. If the two vehicles are not interlocked, the release of this energy with will momentarily 'spring apart' the two vehicles.

(e) The frictional force acting on the target vehicle will slow it such that the two vehicles have a second collision. After this collision, the striking vehicle and the target vehicle will once again have same velocity.

Stage-4: Post-Collision, Sliding to a Stop

(f) The two vehicles slide against the tire friction until they are stationary. All energy has
   been dissipated from the event; either through crush, elastic deformation, or tire wear,
   etc.

Behavior, Detailed

Stage-1: Stationary Crush
Material fracture and crush will occur in the target vehicle once the impulse force is
applied. The target vehicle will slide relative to the ground once the impulse F·t product
is sufficient to accelerate the target vehicle mass and break the frictional force between
the tire and ground. Crush will continue until there is no relative velocity between the
two vehicles.

The amount of crush in the target vehicle will be dependent on;

- the impulse magnitude required to change the target vehicle's momentum (Eq. 4.98, a
  function of mass)
- the strength of the loadpath within the target vehicle between the impact location and
  its significant masses ($s_T(\delta_{T_p})$),
- the stiffness of the loadpath within the target vehicle between the impact location and
  its significant masses (K_{Tm}),
- the intensity of the motion resisting frictional force seen by the target vehicle's tires
  (f),
- the stiffness of the loadpath within the target vehicle between the impact location and
  the target vehicle's tire patch (K_{Tf}), and
- the strength and depth of the deformable portion of the striking vehicle ($s_T(\delta_{S_p})$).

Target Vehicle Strength ($S_T(\delta_{T_p})$)
   With regards to the relationship between crush magnitude and the target vehicle
strength, recall the lesson of Sect. 3.2.3.3; a weaker target object will limit the impulse
force magnitude and will require a longer impulse time duration to achieve the necessary
impulse magnitude to accelerate the object. Figure 4.113 illustrates the engaged loadpaths. The width of the black bars represents each loadpath's relative strength; the *rocker/
floor-bar* loadpath being the strongest, the *side-door* being the weakest.

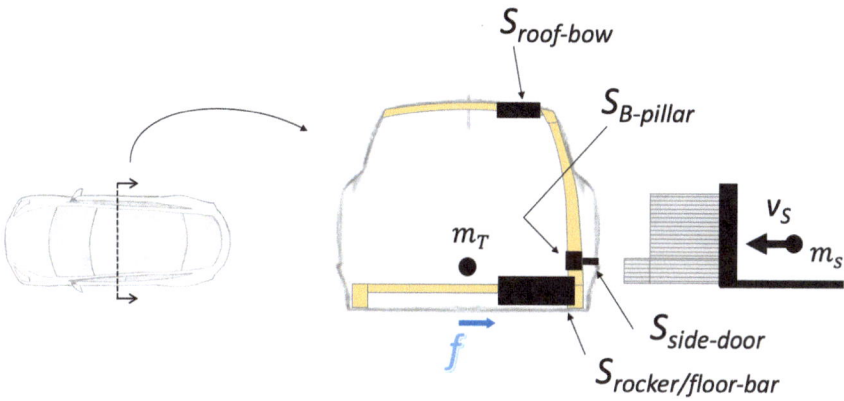

Fig. 4.113 Side-MDB loadpaths and their strengths

Consider the structural performance of the 2022 Malibu (Link 4.24) and the 2022 Outback (Link 4.25) as an example of how the strength of engaged loadpaths and intrusion are related.

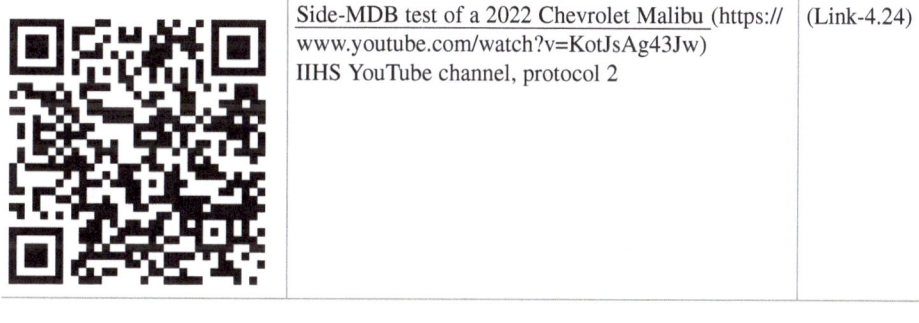

	Side-MDB test of a 2022 Chevrolet Malibu (https://www.youtube.com/watch?v=KotJsAg43Jw) IIHS YouTube channel, protocol 2	(Link-4.24)

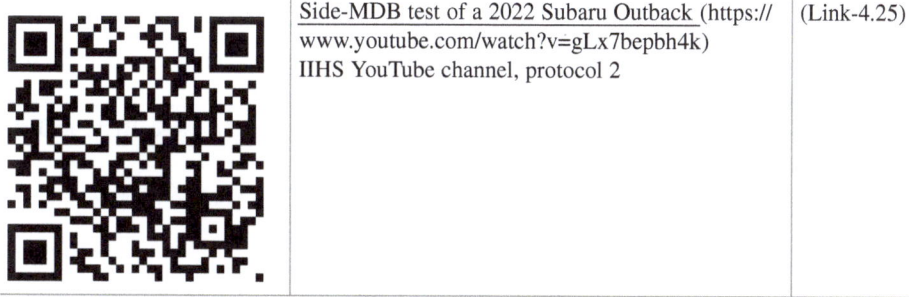

	Side-MDB test of a 2022 Subaru Outback (https://www.youtube.com/watch?v=gLx7bepbh4k) IIHS YouTube channel, protocol 2	(Link-4.25)

In the case of the Malibu, the barrier ''crabbed over' the rocker and engaged the B-pillar. The Outback's rocker is higher than the Malibu's and we find that the barrier

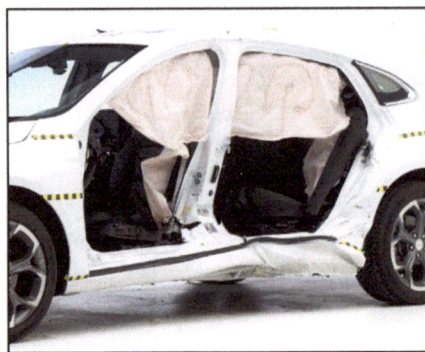

Fig. 4.114 IIHS side-MDB examples, intrusion versus rocker engagement (IIHS)

maintained rocker engagement in the Outback test. The B-pillar intrusions were approximately 270 and 160 mm for the Malibu and Outback, respectively. The condition of their body structure post-test is shown in Fig. 4.114.

It is important to note that achieving good safety ratings with a low height automobile is possible, it is just more difficult. The timing of design development relative to the introduction of crashworthiness loadcases is an important consideration when judging the Malibu's performance, as the vehicle was developed prior to 2015 and is shown here assessed using an IIHS test protocol updated in 2021.

Mass Loadpath Stiffness (K_{Tm})

The stiffness between the impact location and the target vehicle's masses rarely arises as an issue in Side-MDB performance development. The variable is noted here primarily for completeness; understand that it is an element of behavior and keep it in your mental model.

Frictional Force (F)

The frictional force resisting target vehicle motion is dependent on the weight of the target vehicle, the friction coefficient of the tire/ground condition, and the number of tires contacting the ground -but we'll assume that is four.

To put the frictional force into context, consider a small car, a large BEV, and the equations for normal force (N) and frictional force (fs).

Using the equations for normal force and frictional force and vehicle parameters…

$$N = m_T g \tag{4.104}$$

$$f_s = (\mu_s N)/2 \tag{4.105}$$

where:

N	=	normal force
m_T	=	mass of the target object
g	=	acceleration of gravity
f_s	=	static frictional force
μ_s	=	static friction coefficient

Table 4.3 shows us that the frictional force can vary quite significantly depending on the weight of the target vehicle.

Table 4.3 Frictional force examples

	2023 fiat 500×	2022 GMC hummer EV
Curb weight	1500 kg (3305 lb)	4111 kg (9063 lb)
N	15 KN	40 KN
μ_s	0.7	0.7
f_s at each tire	2.6 KN	7.1 KN

Friction Loadpath Stiffness (K_{Tf})

Recall from Sect. 3.2.3.2 that the stiffness of a loadpath will influence how much force can be transmitted thru it. The extremes of this loadpath stiffness can be illustrated by comparing a BOF automobile with large profile tires to a BFI automobile with low profile tires and recognizing the spring rate difference that exist; as shown in Fig. 4.115.

In practice, stiffness of this loadpath is not necessarily a significant focus of design development; however it is important to understand its influence on the target vehicle behavior.

Stage-2: Crush with Target Vehicle Velocity
The second stage begins once the impulse force is sufficient to change the momentum of the target vehicle and overcome the frictional force. The collision event is over once the striking and target vehicles are traveling together at the same speed. Their velocity can be estimated using the momentum equation developed in Sect. 3.2.3.2 and replicated here below as Eq. 4.106. Note that the actual velocity should be lower than this calculated estimate, since this equation does not include the effect of tire friction.

$$v_f \approx v_{So} \frac{2m_S}{m_S + m_T} \tag{4.106}$$

where no friction acts on the system

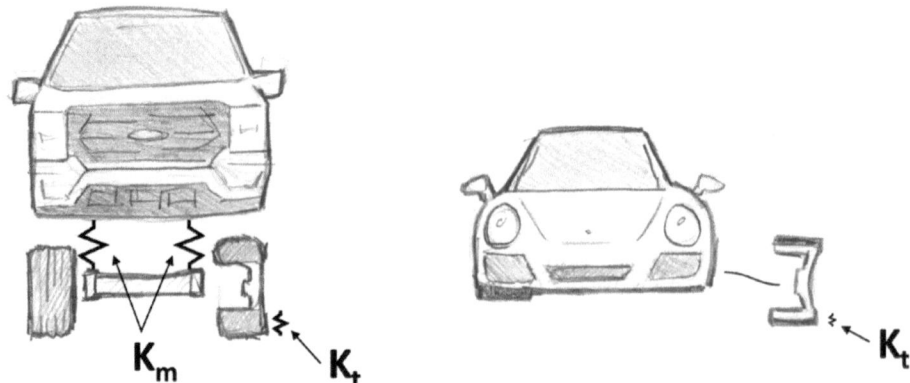

Fig. 4.115 Friction loadpath extremes

Stage-3: Post-Collision, Rebound and Secondary Collision (Optional)
If the two vehicles are not interlocked due to the geometric crush, they will momentarily spring apart due to the release of energy associated with elastic compression. This separation is more likely if the target vehicle has been struck in a relatively stiff location of the stricture; recall from the Malibu/Outback comparison that the Outback exhibited separation while the Malibu did not.

The striking vehicle will collide with the target vehicle once again, since the two vehicles still have momentum but only the target vehicle's motion is being resisted by the friction. Note that the Stage-3 Rebound section of the Rear-MDB loadcase explores the mechanism of energy storage within the test vehicle in detail and is worth referencing.

Stage-4: Post-Collision, Sliding to a Stop
All the residual kinetic energy is eventually expended through the frictional force between the tire and ground.

Numerical computation of this stage has little importance from an engineering design development perspective, but we can model it simply for the sake of completeness. The equation model for stage-4 is based on slowing the remaining kinetic energy by the frictional force.

Starting with a constant acceleration motion equation:

$$v_e^2 = v_f^2 + 2a(d_4) \tag{4.107}$$

$$0 = v_f^2 + 2a(d_4) \tag{4.108}$$

$$d_4 = -\frac{v_f^2}{2a} \tag{4.109}$$

where:

v_f = velocity of the vehicles at the end of the collision
v_e = velocity of the vehicles at the end of the event
d_4 = distance traveled by the vehicles during stage-4
a = rate of deceleration

Some manipulation of Newton's second law....

$$F = ma \qquad (4.110)$$

$$-f = (m_S + m_T)a \qquad (4.111)$$

...substituting N/g for mass, per Eq. 4.104...

$$-f = \frac{N}{g}a \qquad (4.112)$$

...solving for acceleration...

$$a = -\frac{fg}{N} \qquad (4.113)$$

...substituting Eq. 4.114 into Eq. 4.113...

$$\mu_d = \frac{f}{N} \qquad (4.114)$$

$$a = -g\mu_d \qquad (4.115)$$

...finally, substituting Eq. 4.115 into Eq. 4.109 yields:

$$d_4 = \frac{v_f^2}{2g\mu_d} \qquad (4.116)$$

where:

N = normal force
d_4 = distance traveled by the vehicles in stage-4
v_f = velocity of the vehicles at the end of the collision
g = acceleration of gravity
μ_d = dynamic friction coef. between target vehicle tires and ground

Equation 4.116 shows that, as we would hope to find, the distance traveled in stage-4 is dependent only on the velocity of the combined vehicles at the beginning of stage-4, the friction coefficient, and what planet the event takes place on.

4.5.3   Structural Loadpath Topology

There are three loadpaths of interest for Side MBD; the side-door system, the *rocker*, and the *B-pillar*. Each are detailed here, in the order of their event participation. For completeness, the loadpath between the vehicle side structure and the tires will also be covered.

Side-Door System
The first loadpath encountered by the striking vehicle is the side door system, and particularly, the reinforcements that lay within (Fig. 4.116). Longitudinal *door-beams* are engineered to direct load from the barrier to the body structure through the *hinges, latch,* and sometimes interlock features. The *beams* themselves are constructed of a high strength material such as to maintain load dispersion capability as long as possible and to maximize energy absorption. The *hinge* mounting surface of the *side-door-inner-panel* also participate in this loadpath. *Hinge* and *latch* strength are also important, such as to maintain the loadpath.

Rocker
Vertical overlap between the barrier and the *rocker* is required for the *rocker* to participate in the loadpath. Overlap is dependent on what test protocol is being considered and the height location of the *rocker* with respect to the ground.

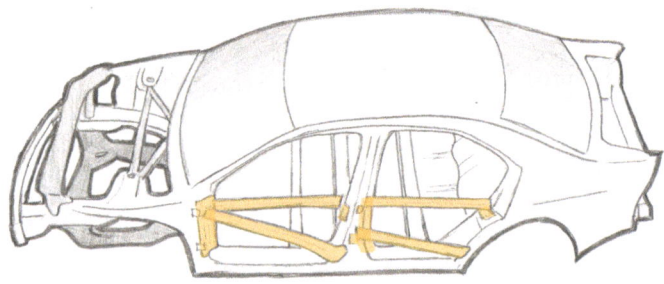

Fig. 4.116   Typical side MDB structural loadpath topology, side door system

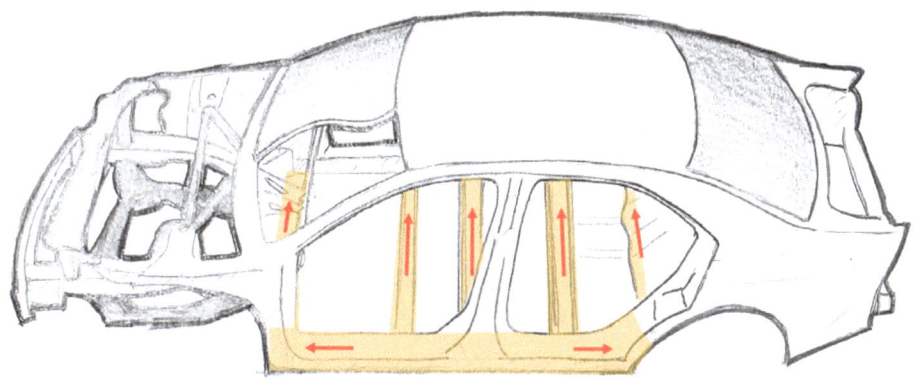

Fig. 4.117 Typical side MDB structural loadpath topology, body

Between protocols, the distance between the bottom of the barrier and the ground ranges from 300 mm (for tests including FMVSS214, EuroNCAP, and UNR95) to 350 mm (for test such as the IIHS Side Crash). Thus, a vehicle tested under a protocol specifying a 300 mm attribute is more likely to have engagement with the rocker.

Different types of vehicles have different rocker heights, largely based on the vehicle's ride height/ ground clearance. A 'low as can be' sports car is unlikely to have rocker engagement under any test protocol while a SUV is likely to have overlap in all. An automobile with the ground clearance of a typical sedan tends to have rocker engagement in lower barrier placement protocols and lack engagement when the barrier is positioned higher.

When barrier engagement with the rocker can be maintained, the impulse force can be dispersed through the rocker to the body center-compartment and floor-bars.

The Rocker and it's loadpath into the vehicle are illustrated in Fig. 4.117.

B-Pillar

Loads are initially applied to the *B-pillar* through the side-door system but eventually the striking vehicle encounters the *B-pillar* directly. The loads from the striking vehicle create a local FDB; like a simply supported beam where the beam constraints are the *roof-side-rail* at one end and the *rocker* at the other, as illustrated in Fig. 4.118a. A *B-pillar* design that has not been engineered for the Side-MDB loadcase tends to buckle mid-height, near the beltline of the vehicle (Fig. 4.118a). Recalling that occupant protection is the ultimate performance measure for this loadcase, it should be easy to see that this deformed shape is undesirable; the intrusion magnitude is high and the size of the structure interfacing the occupant is small. The strength profile of a B-pillar is engineered such that the intrusion magnitude is lower, and a friendly surface is 'presented' to the occupant. Notice that the engineered deformed shape shown in Fig. 4.118b presents a large surface to the occupant; thus distributing the forces involved in the collision between the occupant and the vehicle.

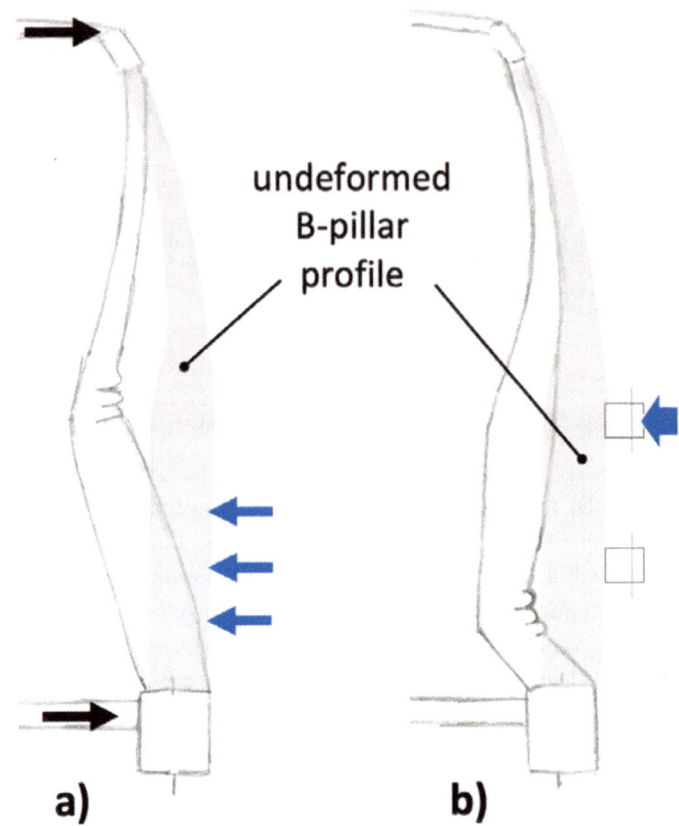

undeformed
B-pillar
profile

a)   b)

Fig. 4.118 B-pillar deformed shapes

Note that there will be more structural demand placed on the B-pillar if the barrier does not engage the rocker or engages the rocker for only a short amount of time.

It is relevant to discuss the timing of the striking vehicle contact with the B-pillar. If the barrier engages the side-door-upper-hinge (Fig. 4.118b, blue arrow) during stage-1 stationary crush, the locally applied force will be large and have a greater tendency to instigate the deformed shape shown in Fig. 4.118a. If, however, the barrier engages the upper-hinge during stage-2, when the vehicle velocities are more similar, the force magnitude will be lower. Adjusting the timing of this engagement is not necessarily possible through vehicle design modification however it can be helpful to inspect the loadpaths for softness and weakness and compare engagement timing with simulation of other vehicles.

Loadpath between the vehicle side structure and the tires
The loadpath between the impulse force and the tires is created simply by following a path to the tires. Such a loadpath includes the structure, and suspension linkages. A typical

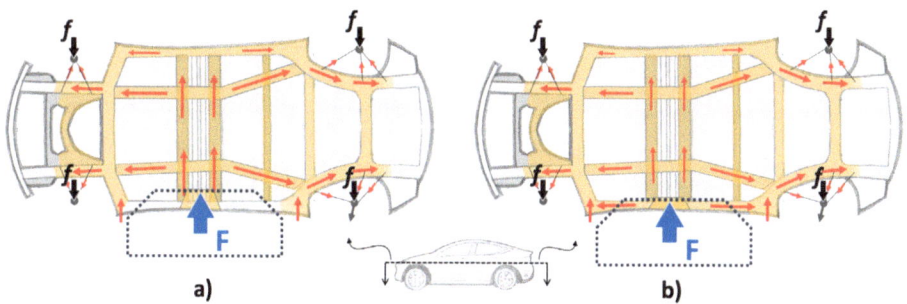

Fig. 4.119 Loadpath to tires

loadpath is illustrated here in Fig. 4.119. Since the elements of this loadpath must be stiff and strong enough to satisfy other loadcases, is rare that weakness in this loadpath cause issue for Side-MDB performance. It is, however, good to recognize its role.

4.5.4   Peculiarities of EVs

Post-Crash High Voltage Safety
Recall the overview of FMVSS305 and UNIECE regulations provided in this chapter's introduction. These regulations apply to vehicles with high voltage content and specify post-crash performance related to; protection from electrical shock, level of electrolyte spillage, and retention of the propulsion battery housing. Both of these regulations apply to the Side-MDB loadcase.

These post-crash high voltage safety regulations become particularly important when high voltage content is near or within the loadpath; vehicle configurations which include propulsion battery content within the tunnel, for example. BEVs with a high ride height, such that the barrier has meaningful engagement with the rocker, might also find that packaging and detail design of an underfloor PBH requires careful consideration with regards to these regulations.

Mass
Simply put, EVs in the era of 600 Wh/l battery cell technology and a 300 mile range requirement are heavier than their ICE counterparts. Equation 4.98 tells us that the impulse magnitude increases as the mass of the target vehicle increases and that the friction force resisting the target vehicle's translation will be higher. Equation 4.103 indicates that the force·intrusion product will increase as the target vehicle mass increases. Ultimately, there will be an increased strength demand on the *door beams,* hinging system, latching system, and Body *B-pillar* to achieve comparable performance as vehicle mass increases.

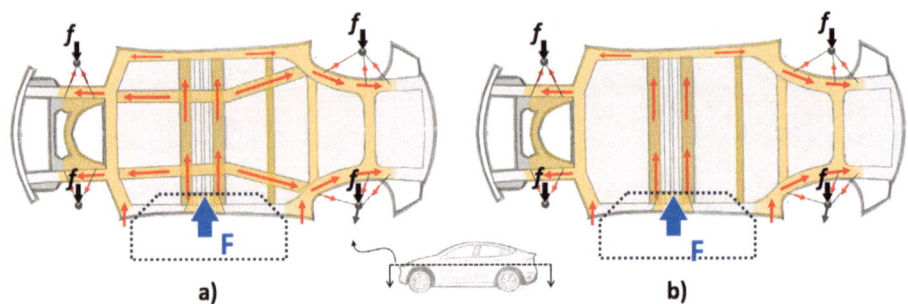

Fig. 4.120 Loadpath from barrier to tire patch; ICE and BEV

Topology

The loss of the traditional center-compartment body longitudinal loadpath that accompanies the full-width underfloor propulsion battery BEV configuration (Fig. 4.120b) can affect the Side-Pole performance since the loadpath between the barrier contact and the tire patch can be weakened and/or softened. Careful thought of the structural topology within the battery housing is one way to address this issue. The concepts described in the Side-Pole 'Peculiarities of EVs' section are also relevant here.

4.6   Rear Moving Deformable Barrier

There are several Rear Moving Deformable Barrier (MDB) loadcases around the world, varying in speed and overlap. The most challenging loadcase is the FMVSS-301 condition, as it has the highest speed and the smallest overlap. FMVSS-301 is a regulatory loadcase in which a test vehicle is struck with side and rear barriers, in independent tests. The purpose of this regulation is to ensure a level of safety with regards to post crash fire caused by fuel leakage. The qualifier "R" is often used to signify the rear barrier test condition of the FMVSS-301 regulation; FMVSS-301R.

The FMVSS-301R test condition can be particularly challenging for two reasons; the speed of the striking vehicle is quite high (80kph, or 50mph), and that only a portion of the target vehicle is struck (70% of its width). The general configuration of the test is illustrated in Fig. 4.121 and the general collision model is shown in Fig. 4.122. Link 4.26 provides an example video of a Rear-ODB test.

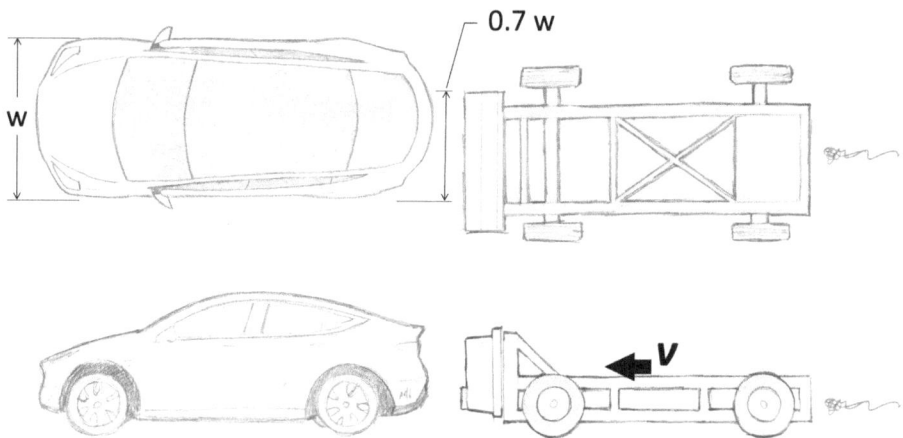

Fig. 4.121 General rear-ODB test condition

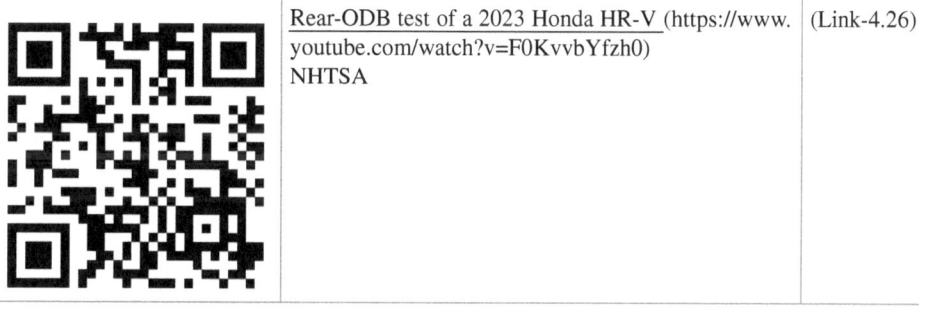	Rear-ODB test of a 2023 Honda HR-V (https://www. youtube.com/watch?v=F0KvvbYfzh0) NHTSA	(Link-4.26)

Fig. 4.122 variable definitions:

s_S	=	is the strength profile of the striking vehicle,
δ_{Sp}	=	is the direction of striking vehicle plastic deformation measure,
δ_{Se}	=	is the direction of striking vehicle elastic deflection measure,
K_{Sm}	=	is the stiffness of the striking vehicle between impulse & mass center,
s_T	=	is the strength profile of the target vehicle,
δ_{Tp}	=	is the direction of target vehicle plastic deformation measure,
δ_{Te}	=	is the direction of target vehicle elastic deflection measure,
K_{Tm}	=	is the stiffness of the target vehicle between impulse & mass center

Within the industry, this test condition is sometimes referred to as "Rear-ODB", as the barrier is both Offset and Deformable.

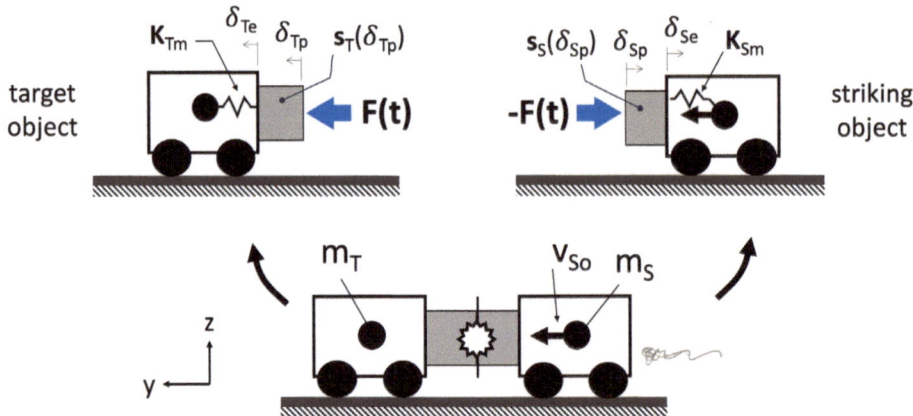

Fig. 4.122 General rear-MDB collision model

4.6.1   Loadcase Modeling

- Plasticity (s_T, s_S): Plasticity exists in both the striking and target objects.
- Friction: As both objects are free to roll in the direction of the collision, any friction that occurs is insignificant and can be ignored.
- Elasticity (K_{Tm}, K_{Sm}): Elasticity exists in both the striking and target objects. In both cases, it is the stiffness between the applied impulse force and the vehicle's generalized mass center that is applicable.

Insight from Governing Equations

Recall from Sect. 3.2.3.1 that impulse is defined as the change in a collision object's momentum (4.117) and that impulse is also equivalent to the area under the F(t) curve (4.118).

$$\text{Impulse}: \boxed{J = p_o - p_f} \qquad [\text{N s}] \text{ or } [\text{kg m/s}] \tag{4.117}$$

$$\text{Impulse}: \boxed{J = \int_o^f F(t)\,dt} \tag{4.118}$$

or

$$\boxed{J = F_{ave}\,\Delta t} \tag{4.119}$$

where:

J =   Impulse

p =   momentum of a collision object

F =   impulse force

Δt =   time duration of collision

...equating 4.117 and 4.118...

$$J = J \tag{4.120}$$

$$F_{ave}\,\Delta t = 1p \tag{4.121}$$

$$F_{ave}\,\Delta t = m_T v_{Tf} - m_T v_{To} \tag{4.122}$$

...recognizing that the target vehicle is stationary at the start of the event...

$$F_{ave}\,\Delta t = m_T v_{Tf} \tag{4.123}$$

...recall that the final velocity can be determined using the Conservation of Momentum law (Sect. 3.2.3.2)...

$$v_f = \frac{m_S}{m_T + m_S}\,v_{So} \tag{4.124}$$

...substitution results in...

$$F_{ave}\,\Delta t = m_T\left(\frac{m_S}{(m_T + m_S)}\,v_{So}\right) \tag{4.125}$$

where:

F_{ave} =   average impulse force

Δt =   time duration of collision

m_T =   mass of the target vehicle

m_S =   mass of the striking vehicle

v_{To} =   initial velocity of the target vehicle

v_{So} =   initial velocity of the striking vehicle

v_f =   final velocity of the vehicles

Equation 4.125 shows that the impulse will increase as the mass of the target vehicle increases, as plotted in Fig. 4.123.

We can also look at this event from the perspective of the conservation of energy law...

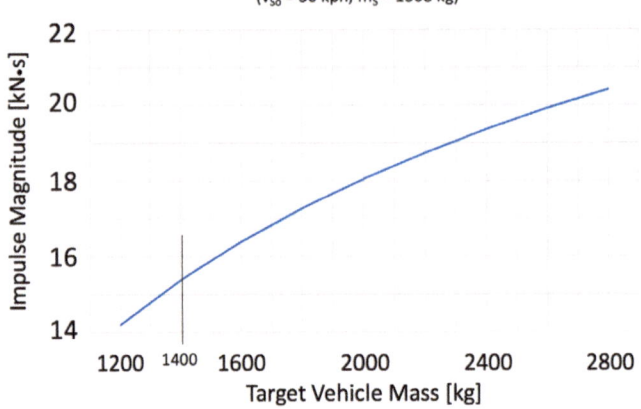

Fig. 4.123 Impulse magnitude versus target vehicle mass

$$E_f = E_o \tag{4.126}$$

$$KE_{Tf} + KE_{Sf} + W_T + W_S = KE_{So} \tag{4.127}$$

$$\frac{1}{2}(m_T + m_S)v_f^2 + F_{ave}\Delta_{Tp} + F_{ave}\Delta_{Sp}$$
$$= \frac{1}{2}m_S v_{So}^2 \tag{4.128}$$

...rearranging results in an equation in an intuitive arrangement; the work done through crush is equivalent to the difference between the initial and final energy...

$$F_{ave}\Delta_{Tp} + F_{ave}\Delta_{Sp} = \frac{1}{2}m_S v_{So}^2 - \frac{1}{2}(m_T + m_S)v_f^2 \tag{4.129}$$

...substituting Eq. 4.124 into 4.129 and some simplification results in...

$$F_{ave}\Delta_{Tp} + F_{ave}\Delta_{Sp} = \frac{1}{2}v_{So}^2\left(m_S - \frac{1}{2}\frac{m_S^2}{(m_T + m_S)}\right) \tag{4.130}$$

where:

F_{ave}	=	average impulse force
Δ_{Tp}	=	crush distance seen within the target vehicle
Δ_{Sp}	=	crush distance seen within the striking vehicle

m_S =   mass of the striking vehicle

m_T =   mass of the target vehicle

v_{So} =   initial velocity of the striking vehicle

v_f =   final velocity of the vehicles

Equation 4.130 shows that some of the striking vehicle's initial kinetic energy is converted into crush in the barrier and crush in the target vehicle. Equation 4.129 reminds us that there is residual kinetic energy at the end of the event.

System Energy Flow
Figure 4.124 illustrates the energy transferer during the collision event. Again, we see that some of the striking vehicle's kinetic energy is converted into crush in the target vehicle ($F_{ave}\Delta_{Tp}$) and into crush in the striking vehicle ($F_{ave}\Delta_{Sp}$) and that the two vehicles have some kinetic energy at the end of the event. Some energy is stored during the event; the amount of which is dependent on the stiffness of the target and striking vehicles (K_{Tm} and K_{Sm}, respectively) and the maximum elastic compression seen in them (Δ_{Te} and Δ_{Se}, respectively). This energy is released once the impulse force is removed and causes the two vehicles to 'spring' apart.

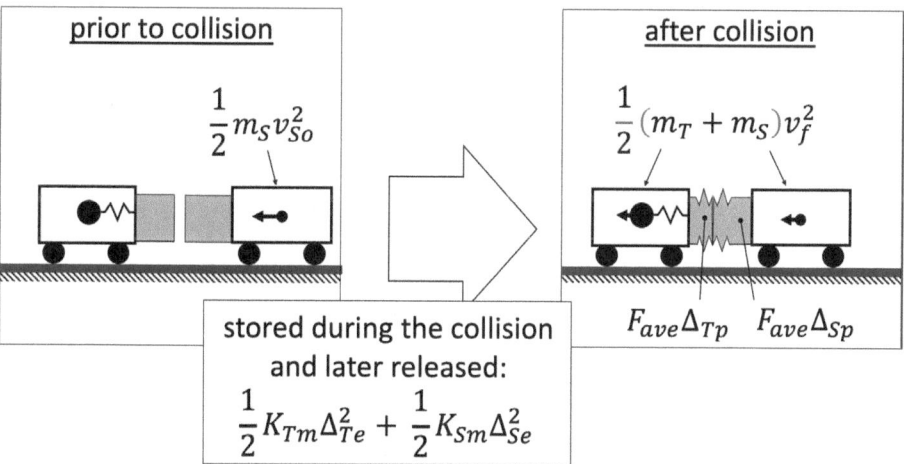

Fig. 4.124   Rear-MDB energy conservation

4.6.2 Fundamental Behavior and Physics

Figure 4.125 illustrates the FBD of the target vehicle. The collision force is resisted only by the target vehicle mass, whose momentum will be changed during the event.

Behavior of the striking vehicle is similar to that in Side-MDB, with the following notable exceptions:

- Side-MDB involves friction between the target vehicle and ground while the target vehicle is free to roll in Rear-MDB.
- the amount of crush seen by the target vehicle tends to be smaller in Side-MDB.

Event Milestones
Figure 4.126 dissects the general behavior of the two vehicles in the Rear-MDB.

Behavior, High-Level
Stage-1: Crush and Elastic Compression

(a) The striking vehicle first contacts the stationary target vehicle.
(b) The barrier portion of the striking vehicle and portions of the target vehicle crush. Although difficult to notice, the velocity of the striking vehicle will reduce as the crush dissipates energy. The target vehicle does not typically begin to exhibit notice-able velocity in this stage. Elastic energy is stored within the two vehicles structure

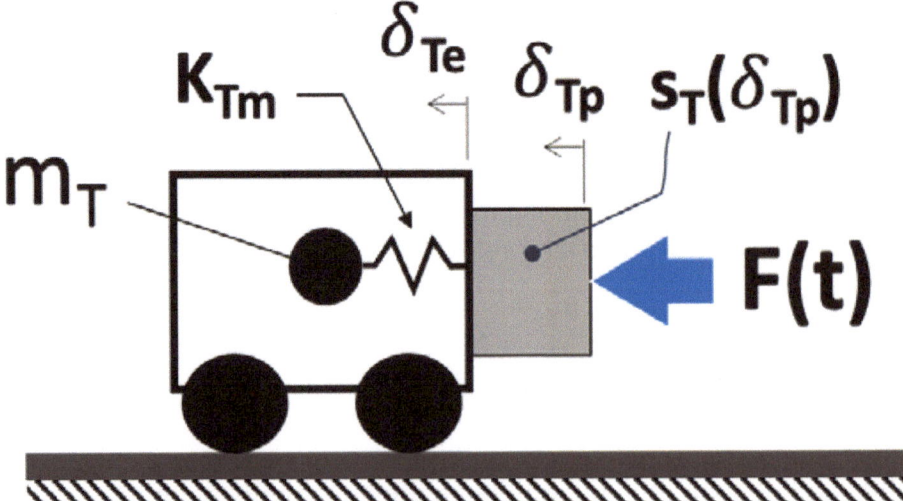

Fig. 4.125 Rear-MDB FBD

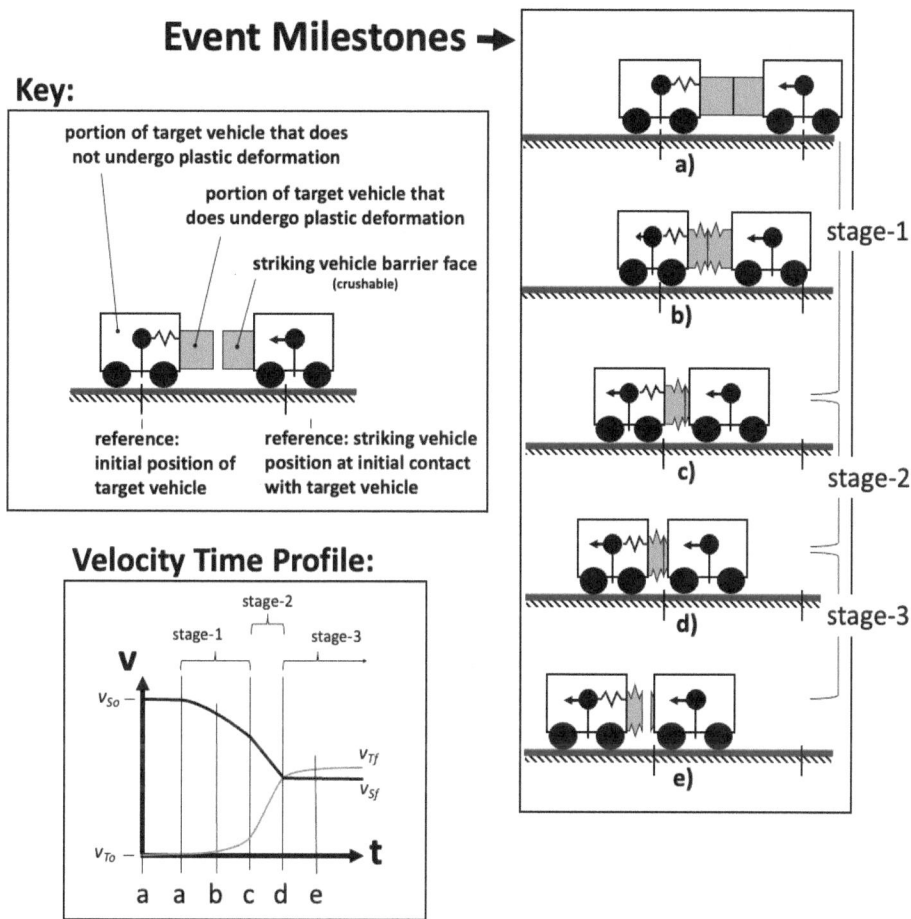

Fig. 4.126 Rear-MDB event milestones

as force is applied by the striking vehicle and the target vehicle's mass 'resists' acceleration, as detailed in Sect. 3.2.3.2.

Stage-2: Stiff MDB—Target Vehicle Acceleration

(c) The barrier on the striking vehicle is fully crushed and the stiffness and strength of the striking vehicle abruptly increases. Crush within the target vehicle continues and its acceleration increases rapidly as the now stiff MDB engages stronger elements of the target vehicle.

Stage-3: Spring-Back and Residual Velocity

(d) The two vehicles have the same velocity. No more crush occurs. The elastic energy stored within the two vehicles is released and the target vehicle 'springs off' of the striking vehicle.
(e) The two vehicles continue to roll forward. -in an actual physical test, braking or capturing methods are employed to stop the vehicles.

Behavior, Detailed

Stage-1: Stationary Crush

Crush occurs in both the target vehicle and the deformable portion of the MDB during this stage. The most significant crush occurs in the target vehicle's *rear impact beam*, the *bumper EA foam*, and the rearward portion of the *body rear-compartment-mid-rails*.

Like that done for the Full-Frontal loadcase, the vehicle structure can be subdivided into segments based on their participation in the crush stage (Fig. 4.127).

- *crush zone*: where crush is to occur
- *back-up structure*: the zone that remains unyielding during the event
- *transition zone*: where lower levels of crush might occur

The force at which each of these segments yield and begin to crush is called its "load capacity". The load capability, or strength, of each segment is engineered to ensure crush occurs in the intended location during the event. The strength of the back-up structure is higher than that of the transition zone and crush zone and the strength of the transition zone is higher than that of the crush zone.

Geometric crush initiation features might be added to the rearward portion of the *rear-compartment-mid-rail* to ensure this strength hierarchy and/or promote efficient crush behavior. Examples of geometric initiation features are illustrated in Fig. 4.128.

The magnitude and location of crush within the target vehicle is depended on the strength of each loadpath segment.

Crush in stage-1 of Rear-MDB will occur in the weakest portion of the target vehicle and the loadpath is engineered to maximize the energy absorption characteristics of this crush zone.

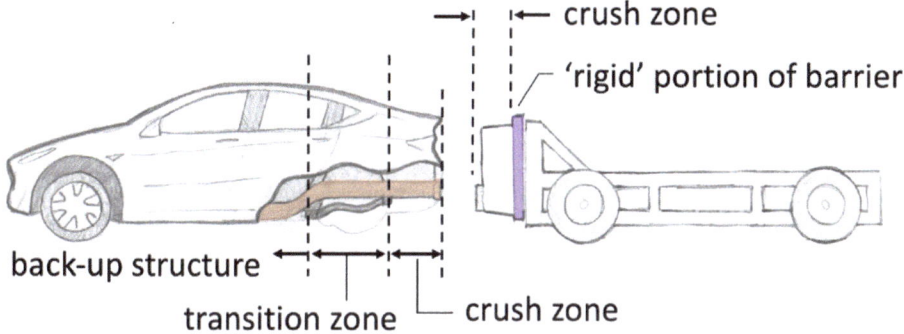

Fig. 4.127 Zones of the rear structure and MDB

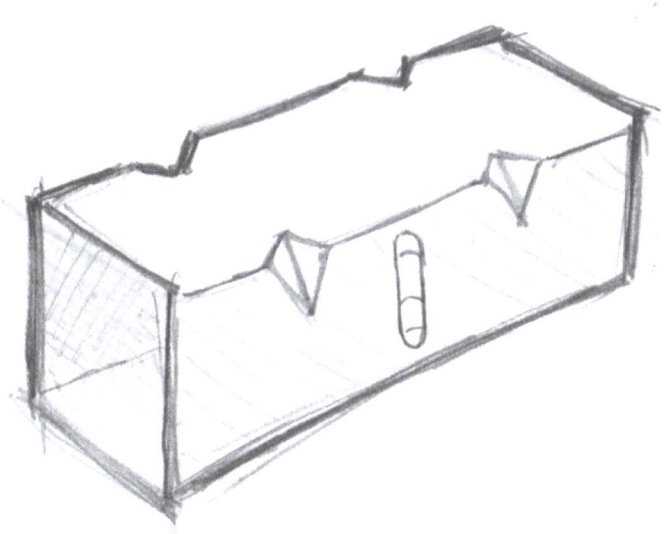

Fig. 4.128 Example of initiation features in a rear-compartment-mid-rail

Although the relative velocity between the two vehicles is high, the impulse force magnitude is not necessarily high. Recall from Sects. 3.2.3.2 and 3.2.3.3 that the impulse force is dependent on the stiffness and strength of the loadpath. Figure 4.129 illustrates the general stage-1 load capacities for the target and striking vehicles, with some foreshadowing of stage-2.

Considering the relatively low strength of the two vehicles, we would expect to see an impulse force like shown in Fig. 4.130 for stage-1 of the Rear-MDB loadcase.

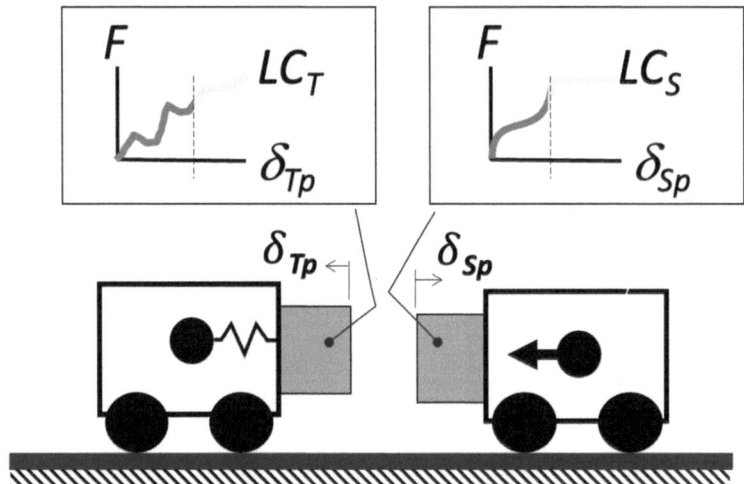

Fig. 4.129   Generalized load capacities, stage-1

Fig. 4.130   Generalized
rear-MDB impulse profile,
stage-1

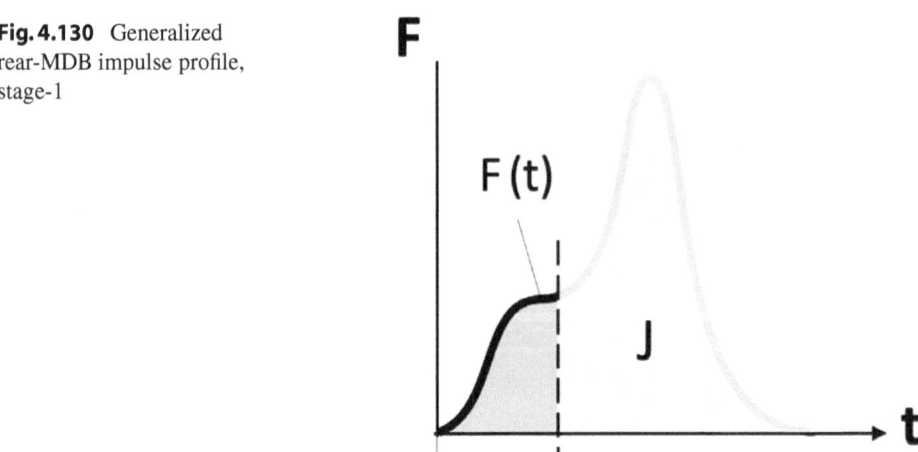

The target vehicle will not begin to move until a sufficient impulse has been applied, so it typically does not necessarily begin to exhibit significant motion in stage-1 of the event.

Stage-2: Stiff MDB—Target Vehicle Acceleration

Stage-2 begins once full crush is achieved in the striking vehicle's deformable barrier. Here, the two vehicles still have significant relative velocity, and the strength of the striking vehicle has quickly and significantly increased. The generalized load capacities shown in Fig. 4.131 illustrate this change.

The impulse force magnitude increases as the stronger striking vehicle encounters stronger elements of the target vehicle structure. We would expect to see an impulse force curve similar to that shown in Fig. 4.132.

The collision event is over once the two vehicles have the same velocity. Without relative velocity there is no longer an impulse force or crush within the vehicles. From a

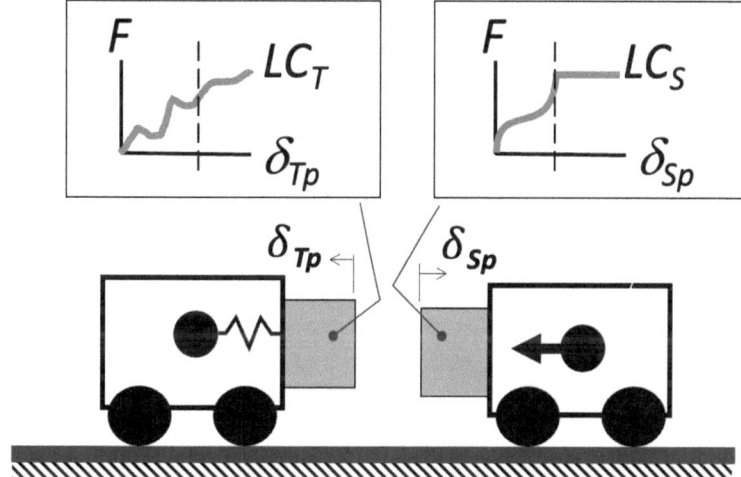

Fig. 4.131 Generalized load capacities

Fig. 4.132 Generalized
rear-MDB impulse profile

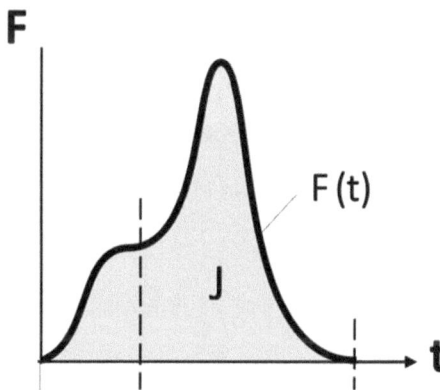

structural performance standpoint, the event is over; no more damage can be done to the target vehicle.

Stage-3: Spring-Back and Residual Velocity
The energy stored in the two vehicles during the event, as detailed in Sect. 3.2.3.2, is released as the impulse force is removed. The release od this energy is evident in how the two vehicles 'spring apart' after collision.

The two vehicles have a residual velocity at the end of test, which is irrelevant to the performance of the vehicle under test; its plastic deformation having already stopped.

4.6.3   Structural Loadpath Topology

The crush zone portion of Rear-MDB's loadpath topology includes; the *rear-impact-beam* and the near-side body *rear-compartment-mid-rail*. The portion of the loadpath responsible for transmitting the applied barrier force to the vehicle masses might include several different paths and components, as illustrated in Fig. 4.133.

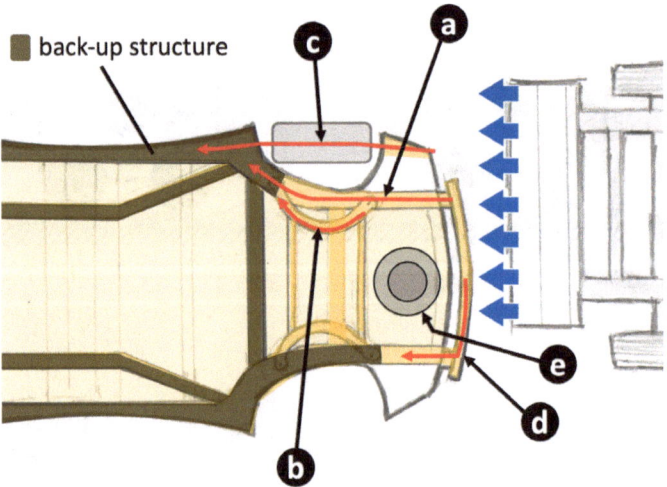

Fig. 4.133 Typical rear-MDB structural loadpath topology

These paths can include;

a. the near-side body *rear-compartment-mid-rail* to *torque-box*
b. the *rear-cradle* to *torque-box*
c. the *bootleg-pocket* to *rear-wheel* to *rocker*
d. the *rear-impact-beam* to far-side body *rear-compartment-mid-rail*.
e. the *spare-tire*, if present.

Usage of the rear wheel loadpath can be seen in the following test of a Ford Focus: We can tell that the Focus's near-side rear wheel is under compression as it does not rotate as it translates forward.

| | Rear-ODB test of a 2012 Ford Focus Electric (https:// www.youtube.com/watch?v=UzBWK7VhuFU) NHTSA | (Link-4.27) |

Leveraging the far-side structure through the *rear-impact beam* requires good joint strength between the *impact-beam* and the body structure. When leveraged, the far-side *rear-compartment-mid-rail* will absorb energy by bending inward and/or exhibiting some amount of crush.

The *rear-torque-box* is an important part of the back-up structure and it is important to recognize the challenge associated with engineering its strength. There are two *torque-box* geometric discontinuities that are fundamental to automotive design, as illustrated in Fig. 4.134;

a. the vertical offset between the *rear-compartment-mid-rail* and the *rocker* due to the demand for a low floor in the occupant compartment and accommodation for rear suspension link articulation below the *rear-compartment-mid-rail*.
b. the lateral offset between the *rear-compartment-mid-rail* and the *rocker* due to the demand for a wide occupant compartment and packaging of the rear wheel.

Recall from the conservation of Energy work above that there is a relationship between the dissipated kinetic energy, the average impulse force, and the crush occurring in the two vehicles, written here in Eq. 4.131.

$$\left(KE_o - KE_f\right) = F_{ave}\Delta_{Tp} + F_{ave}\Delta_{Sp} \qquad (4.131)$$

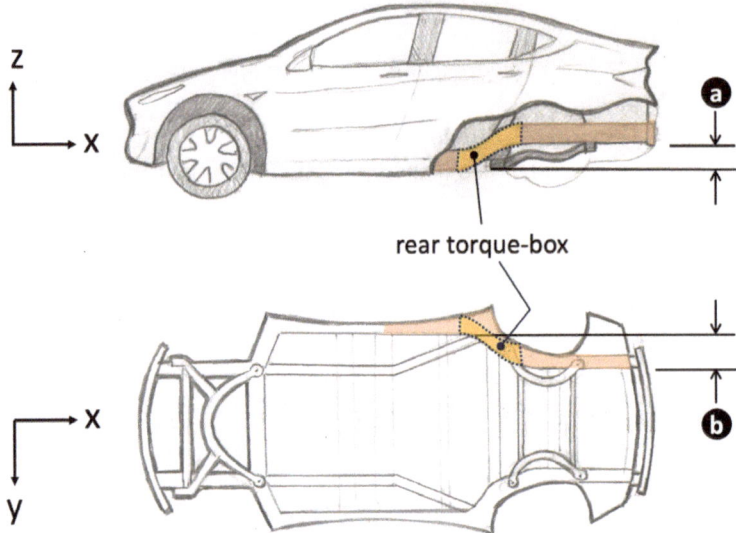

rear torque-box

Fig. 4.134 Fundamental loadpath discontinuities

Assuming that full crush occurs in the striking barrier, we can extrapolate Eq. 4.131, which shows proportionality between the average force applied to the back-up structure (F_{ave}) and the crush distance (Δ_{Tp}).

$$\boxed{F_{ave}\, \Delta_{Tp} \propto KE_o - KE_f} \qquad (4.132)$$

This relationship can be leveraged during the design development process, considering the variables illustrated in Fig. 4.135; the strength of the back-up structure and the crush zone length. Assuming a fixed impulse force magnitude, the strength of the back-up structure needs to increase as the crush zone decreases in length, or more specifically, as less energy is absorbed by the overall crush zone. The relationship be useful in early vehicle development to estimate the required crush zone length based on the vehicle mass and a known range of feasible back-up structure strength.

The relationship can also be used to develop efficiency; what is the optimal balance between crush zone length and back-up structure strength with regards to minimizing vehicle cost and mass?

Geometric discontinuities in the rear-compartment-mid-rail can limit the energy absorption capabilities of the crush zone and minimizing them can require a design challenge to other vehicle systems. For example, in some cases, the rear suspension spring and shock position can conflict with the rail topology, threatening local section size reductions as illustrated in Fig. 4.136.

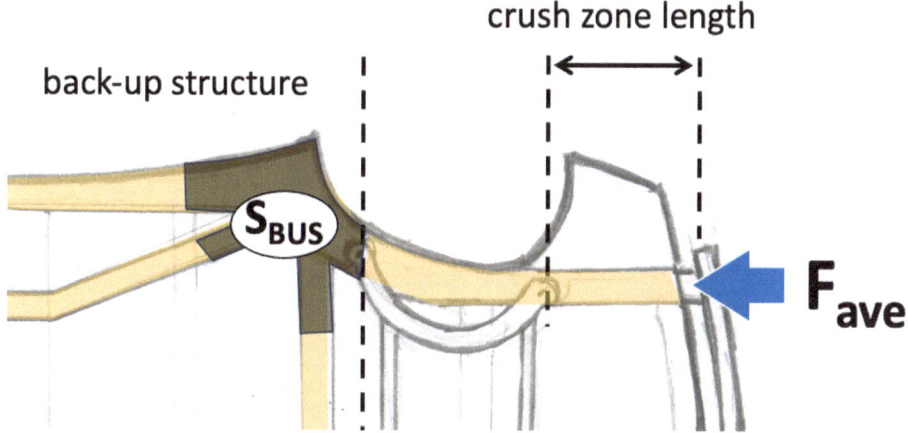

Fig. 4.135 Back-up structure strength and crush zone length

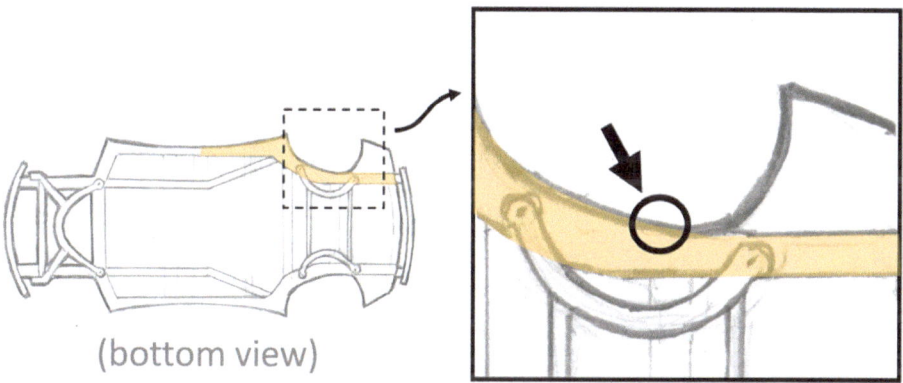

Fig. 4.136 Potential rear suspension shock and spring influences

Alignment of Topology

The FMVSS310R barrier has two different strength zones, a higher strength lower zone and a lower strength upper zone, as illustrated in Fig. 4.137. Alignment of a vehicle's structural topology to the stronger zone is advantageous for overall energy absorption and positioning of the vehicle's structure is considered during the vehicle development process, as much as the vehicle's ride height and other *rear-impact-beam* positioning influencers will allow.

Adjusting a loadpath to align to a test barrier could sound like 'designing to the test', however consider that FMVSS301R is a regulatory test; it must be passed for the vehicle to be sold in the United States. You could petition NHSTA to change the test protocol

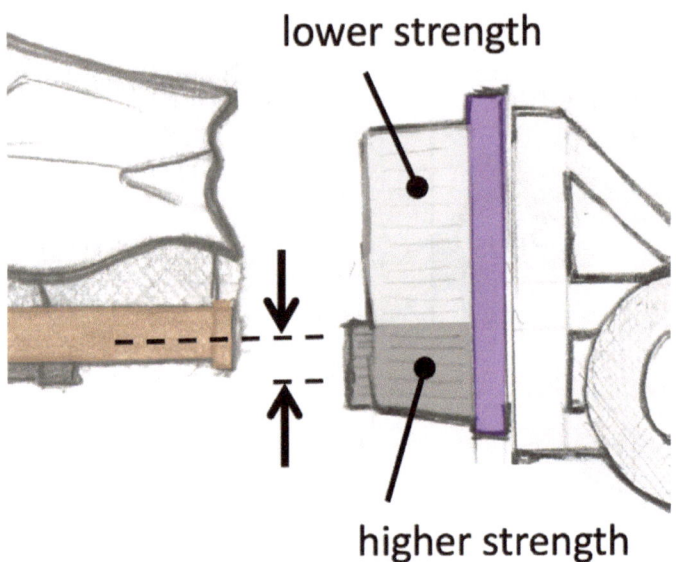

Fig. 4.137 Loadpath alignment to barrier

if you possessed data showing that a different *rear-impact-beam* height is safer, but your vehicle must pass FMVSS301R in the interim.

4.6.4 Peculiarities of EVs

Post-Crash High Voltage Safety
Recall the overview of FMVSS305 and UNIECE regulations provided in this chapter's introduction. These regulations apply to vehicles with high voltage content and specify post-crash performance related to; protection from electrical shock, level of electrolyte spillage, and retention of the propulsion battery housing. Both regulations apply to the Rear Barrier loadcase.

Thoughtful packaging of high voltage components and wiring with respect to structural capabilities and behaviors is the most efficient way to engineer for these post-crash high voltage safety regulations; efficient vehicle design solutions are the result of understanding the structural capability and behavior of the rear-MDB crash strategy early in the vehicle development process and packaging sensitive components accordingly.

Mass
BEVs in the era of 600 Wh/l cell chemistry and a 300 mile range requirement can be significantly heavier than their ICE counterparts. Equation 4.125 tells us, and Fig. 4.123 shows us, that the impulse magnitude will increase as the mass of a collision object

increases. Also, Eq. 4.130 shows that the force·intrusion magnitude will increase as the target vehicle mass increases. Hence, the vehicle structure of these EVs must incorporate a larger amount of energy absorption, a stronger back-up structure, or a combination of these things if it intends to match the intrusion performance of a lighter vehicle.

Note that the importance of efficient loadpath and back-up structure topology will increase as the impulse force increases. Engineering an increased force capacity when discontinuities exist, such as illustrated in Fig. 4.136 can require heavy, costly reinforcement. Similarly, *rear-torque-box* strength is increasingly important as mass increases. Efficient geometry in this area, geometry that transitions *rear-compartment-mid-rail* forces through the vertical and lateral offsets, will also be critical to achieving efficient performance.

Non-collision Loadcases

5

Abstract

There are many loadcases that a manufacturer considers during the development of automotive structure that are not crash events, and several of these loadcases often define local or global structural loadpath topology. This chapter covers those loadcases, uncovering the governing physics, outlying structural loadpaths, and explaining how the EV vehicle configuration might alter the structural strategy. Building the reader's intuition of structural behavior is a goal of the chapter, and the material is presented in a manner to foster that. How the EV vehicle configuration impacts loadcases is also captured.

5.1   Interface Stiffness

As introduced in Sect. 1.2, interface stiffness is an assessment of the structure's stiffness at a location where something is attached. These locations tend to be at isolated interfaces, where there is a bushing present, as illustrated in Fig. 5.1.

Noise and Vibration

The term 'isolated' stems from the bushing's purpose in the joint, which is to reduce the amount of vibration energy transmitted through it. The bushing attempts to isolate the structure from the incoming energy; reducing the amount of noise and vibration that transmits through the joint which can eventually be felt by the vehicle occupant(s). An example of a complete 'source, path, receiver' model is illustrated in Fig. 5.3.

The stiffness of both the bushing and the interfacing structure influence the isolation properties of the joint. Isolation improves as the ratio of bushing displacement to total displacement $\delta_{bush}/(\delta_{bush} + \delta_{str})$ increases; the bushing does more of the 'work' with

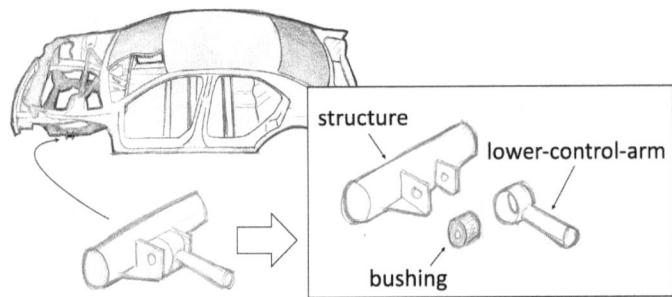

Fig. 5.1 Example of an isolated joint

regards to displacement under applied loads. We can refer to the measure $\delta_{bush}/(\delta_{bush} + \delta_{str})$ as 'joint isolation efficiency'.

Figure 5.2 illustrates how joint isolation efficiency changes as the structural stiffness increase relative to a given bushing stiffness.

When the structure is soft, it has significant participation in the total displacement; joint isolation efficiency is low. As the stiffness of the structure increases, more of the displacement is seen by the bushing and joint isolation efficiency increases. Note that there is a diminishing rate of return; there is a point where increased structural stiffness provides less improvement in joint isolation efficiency. This is the basis of a joint stiffness 'rule of thumb', saying that the structure should be at least ten times as stiff as the bushing.

This 'rule of thumb' has limits, however, as there can be joints in an automobile where the bushing is quite stiff; on the order of 20 kN/mm, perhaps. In these cases, 'ten

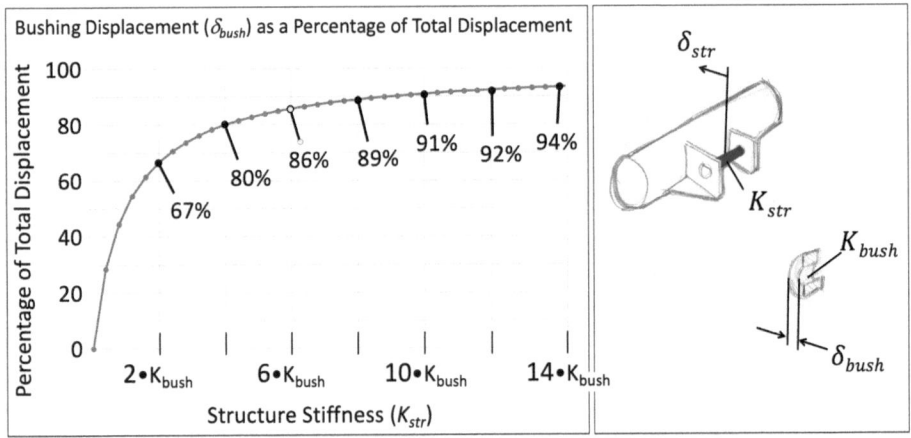

Fig. 5.2 Stiffness ratio rule of thumb

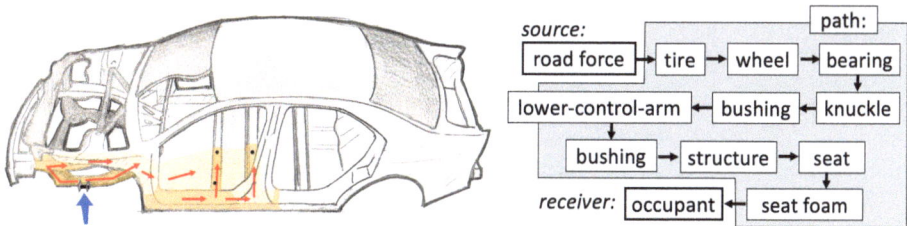

Fig. 5.3 Example of an NV loadpath

times' structural stiffness is impractical. Careful engineering and requirement balancing is applied.

The Big Picture—the Entire Path of Noise or Vibration

It's appropriate to recognize the complete picture of noise or vibration transmission, from the energy source to the occupant. A path can be traced for every source of interest and an example is shown here in Fig. 5.3. Consider an automobile driving in a straight line over rough concrete. The vertical vibrations applied by the road onto the tire travel through all the suspension interfaces as it makes its way to the occupant. The transmission path through the lower-control-arm interface is illustrated in this 'source, path, receiver' model.

In this example, we see that there are many components and a couple of isolated interfaces in the path. The global structure is also in this path and plays a role in the transmission of NV energy, from the lower-control-arm interface to the seat attachments.

The participation of this loadpath can be quantified by measuring the vibration at the seat interface as a function of a force applied at the interface location.

Different sources and paths are more likely to produce undesirable noise or vibration for the occupant(s) and thus more likely to be the focus of engineering development; vibration from the engine is an example.

Ride and Handling

Interface stiffness is also important with regards to achieving the desired dynamic behavior of an automobile; as it travels over bumps and through curves, for example. Ideally, wheel motion should be dictated only by the suspension geometry and the articulation of suspension links. Considering that nothing, including automotive structure, is rigid, wheel control will be influenced by the stiffness of the structure and the setting of interface stiffness targets take this into consideration.

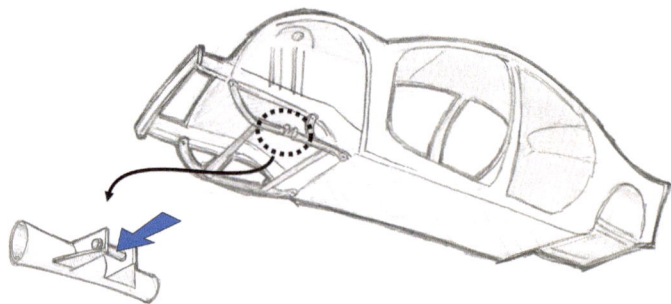

Fig. 5.4 Front suspension lower-control-arm lateral interface stiffness location

Metrics

Two typical metrics exist for Interface Stiffness within the automotive industry, a static stiffness assessment (or one that approximates it) and a dynamic stiffness assessment. The assessment methods are not necessarily consistent throughout the industry so elaborating on the loadcase details is not warranted. It is however noteworthy to say that each have their own advantages; static stiffness measures tend to require less complicated computer simulation models while dynamic stiffness measures can indicate frequencies where the structure has reduced stiffness (due to natural frequencies in the system). The latter becomes relevant in cases where the source vibration occurs at a specific frequency; the cylinder firing process of a 4-cylinder ICE is inherently unbalanced and the engine will produce vibration at a specific frequency while idling, for example.

Relative Influence of Structure

It is interesting to consider the relative participation of structure in Interface Stiffness. Consider front suspension lower-control-arm lateral stiffness, as illustrated in Fig. 5.4.

Certainly, the detail design of the lower-control-arm interface clevis will have a significant effect on the interface stiffness. The lateral stiffness of the cradle side-rails will also come into play, as the applied force will tend to displace the cradle-rail inboard. Even deeper into the structure, we find that the body's resistance to torsional rotation around its mass center will also affect the stiffness measure. It is important to recognize that both the local and global structure will influence Interface Stiffness performance. A relatively soft body will constrain the Interface Stiffness such that no amount of design improvement to the local structure will produce a satisfactory stiffness result. The opposite condition is perhaps more intuitive, where a relatively soft local structure will constrain the Interface Stiffness such that no amount of design improvement to the global structure will produce a satisfactory stiffness result. In practice, computer simulation is used to determine which part of the overall structure should be improved, in cases where Interface Stiffness is underperforming its target.

Typical Values

The appropriate stiffness at any given interface will depend on how the joint functions with regards to motion control or the susceptibility of noise or vibration input. Some examples:

• The front suspension spring interface (Fig. 5.5) on the body plays a significant role in vertical motion control of the spring, as the spring's function is compression about its axis. Thus, the vertical stiffness target at this location on the body will very likely be higher than the fore-aft or lateral targets.

• The front suspension lower-control-arm (Fig. 5.1) functions primarily as a lateral control member and thus its interface on the cradle will require the highest stiffness in the lateral direction; the fore-aft and vertical stiffness targets will be less.

• The powertrain rear mount (typically between the powertrain and the cradle rear cross-member, as illustrated in Fig. 1.7) functions as a vertical motion control element but also isolates the vertical vibrations caused by the running motor. The vertical stiffness target at this location on the cradle will be higher than the fore-aft or lateral targets.

It's perhaps helpful to note that sometimes the 'natural stiffness' of the interface surpasses the stiffness need of the joint. Consider the front suspension shock (spring) mount illustrated in Fig. 5.5 as an example. Based on the shock's function and noise transmission susceptibility, the interface might require a fore-aft stiffness on the order of 8 kN. This location on the body is already much stiffer due to local structural topology and/or stiffness driven by other loadcases. Here, the body at the shock interface might have a 'natural' fore-aft stiffness well over 10 kN because of the longitudinal nature of the *front-compartment-upper-rail* topology and its proximity to the stiff *front-body-hinge-pillar* and *A-pillar*.

Target interface stiffness values for any given vehicle depend on many things. The vehicle's ride and handling and noise and vibration expectations are among the significant factors. Target interface stiffness values will vary from vehicle to vehicle but the following examples can provide some perspective.

Typical interface stiffness values

• cradle or body at front suspension links (X, Y, Z): 8 kN/mm, 15 kN/mm, 8 kN/mm
• cradle at steering rack attachment (X, Y, Z): 5 kN/mm, 18 kN/mm, 5 kN/mm
• body at suspension spring/damper attachment (X, Y, Z): 8 kN/mm, 8 kN/mm, 15 kN/mm
• body at isolated rear cradle attachment (X, Y, Z): 8 kN/mm, 15 kN/mm, 10 kN/mm
• cradle or body at powertrain mount attachment (X, Y, Z): 5 kN/mm, 5 kN/mm, 10 kN/mm

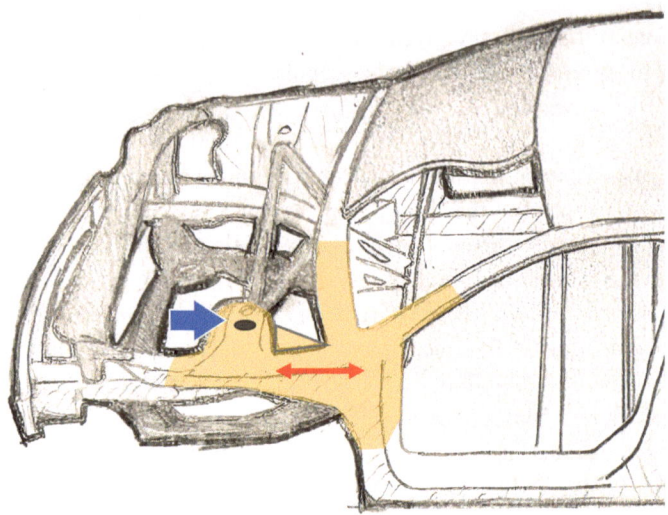

Fig. 5.5 Example of natural stiffness of a structure

5.2   Mounted Component Natural Frequency

The mounting provisions for components attaching to automotive structure must be stiff and strong enough to satisfy local performance expectations. These expectations are largely derived from the desire that the mounted component does not induce vibration which can be felt by the occupant(s). The 12-V battery is a classic example of a mounted component, and it is illustrated in Fig. 5.6.

It is beneficial to understand the concept of a mode map as we consider mounted component vibration. A mode map is a way to collect and visually display noise and vibration source, path, and receiver information. A simplified mode map can be seen in Fig. 5.7.

The first row of the mode map documents noise and vibration sources and the frequency range at which they produce noise and/or vibration; these frequency ranges are sometimes referred to as 'excitation frequencies'. The third row simply indicates that the occupant (the receiver) can perceive noise and vibration in this frequency range. The second row captures the natural frequencies of the 'paths'; those things between the source and the receiver.

There is a general strategy in which the natural frequency of paths and mounted components are engineered and tuned such that they are not excited by sources and that the natural frequency of paths are spaced such that they do not occur at the same frequency. In

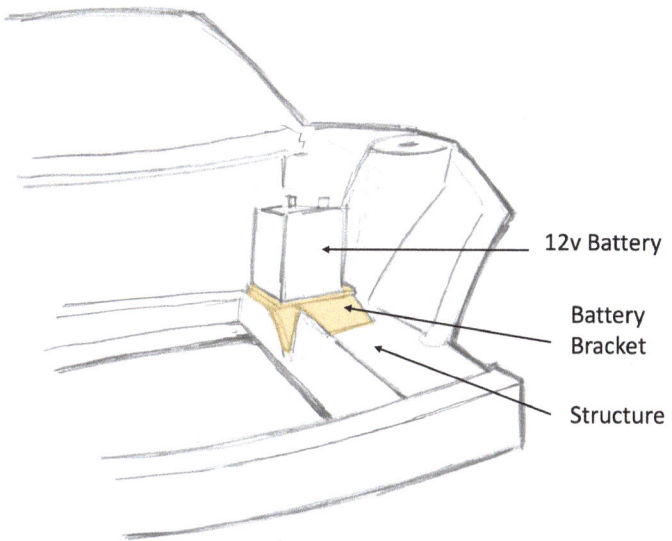

Fig. 5.6 A mounted component example, the 12-V battery

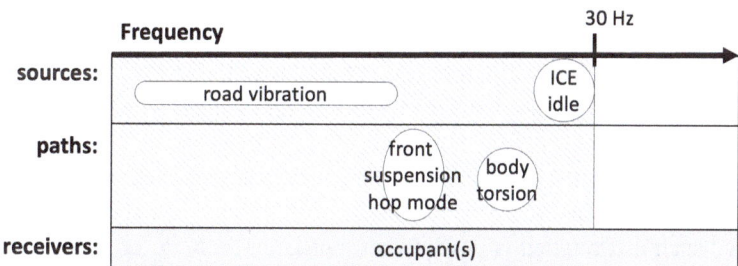

Fig. 5.7 Simplified mode map

Fig. 5.7, notice how the frequency of the ICE idle would not excite the body torsion natural frequency and that the front suspension hop and body torsion natural frequencies are spaced apart within the frequency spectrum.

We see from the simplified mode map that noise and vibration sources tend to occur below 30 Hz. Accordingly, the first natural frequency of components mounting to the structure tends to be targeted to be at 30 Hz or above and stiffness of the attachment strategy and local mounting structure is thereby developed.

Although the primary motivation for placing mounted component natural frequencies above source frequencies is based on occupant satisfaction, it is relevant to note that mounted components excited by a source frequency pose a durability risk. In these cases, the natural frequency of the mounted component would be excited during the vehicle operation, the component would experience vibration regularly, and the local mounting structure would be susceptible to fatigue failure.

5.3   Global Static Torsional Stiffness

Global Static Torsional Stiffness (GSTS) is a subsystem loadcase that assesses the vehicle structure's stiffness under a twisting load applied about the longitudinal axis, as illustrated in Fig. 5.8. The qualifier "Global" signifies that the entire vehicle structure is being assessed.

Vertical forces at the body's suspension spring interfaces are frequently applied during operation of an automobile and these forces often occur asymmetrically; where the force applied to the left does not equal that applied to the right. When such asymmetrical vertical loading occurs, there is an element of torsional loading applied to the vehicle.

Examples of asymmetric vertical loading include; an automobile navigating a curved road, an automobile driving over rough road, and an automobile turning into a ramped driveway. Vehicle level performance such as ride and handling and noise and vibration are affected by global static torsional stiffness.

The entire vehicle structure is considered when assessing GSTS, the content of which is outlined in Sect. 1.3. Because mass does not have a meaningful effect on static stiffness assessments, physical assessments of GSTS can be performed on a completely assembled automobile, as long as the test fixture can get access to the spring interfaces. Note here that assessments are typically performed with the doors open to avoid any contribution that they might provide.

To assess GSTS, motion is constrained at the rear of the structure while force is applied to the front. The locations where the rear suspension springs interface on the structure are constrained such that rotation is possible but translation in X, Y, or Z is not allowed. At the front, an equal and opposite force is applied to the structure at the front suspension spring interfaces. This is illustrated in Fig. 5.9.

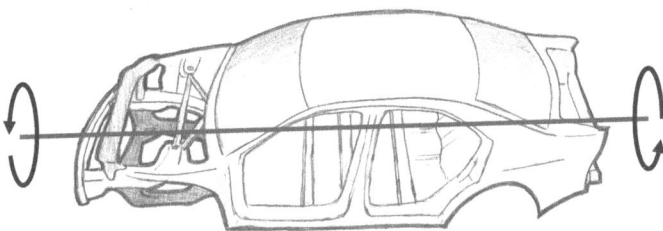

Fig. 5.8 General GSTS loadcase condition

Fig. 5.9 GSTS forces and boundary conditions

5.3.1   Loadcase Modeling

Insight from Governing Equations
A model representing vehicle structure GSTS can be constructed by starting with the equation for angle of twist…

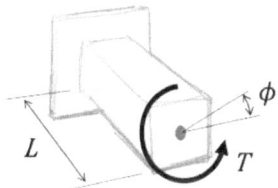

$$\phi = \frac{TL}{JG} \qquad \text{[radians]} \tag{5.1}$$

where:

ϕ	=	angle of twist (radians)
T	=	applied torque (N m)
L	=	length of the object (m)
J	=	polar moment of inertia of the object's cross section
G	=	object's modulus of rigidity

…rearraigning per the definition of stiffness; K = load/displacement…

$$K_t = \frac{T}{\phi} = \frac{JG}{L} \qquad \text{[N m/radians]} \tag{5.2}$$

…converting from radians to degrees yields…

$$\boxed{K_t = \frac{T}{\phi} = \frac{JG}{L}\left(\frac{\pi}{180}\right)} \qquad \text{[N m/degrees]} \qquad (5.3)$$

Equation 5.3 is the standard form of GSTS. However, we can modify the equation further by substituting force and distance for torque, simply to underscore how these variables affect torsional stiffness…

$$K_t = \frac{F}{\phi} = \frac{JG}{wL}\left(\frac{\pi}{180}\right) \qquad \text{[N/degrees]} \qquad (5.4)$$

Here we see that torsional stiffness is a function of vehicle track width (w) and wheelbase (L). Increasing wheelbase or track width will have a negative effect on torsional stiffness. Note that more accurate calculations can be performed using the lateral distance between the suspension spring interfaces (average of front and rear) and the fore/aft distance between front and rear suspension spring interfaces, for w and L respectively. Track width and wheelbase are used simply because these values are typically published, easily found, and provide a good approximation of the theoretically correct measurements.

We also see that torsional stiffness is a function the modulus of elasticity. It is appropriate to visualize this characteristic as a compilation of discrete measures of elasticity between the front and rear suspension spring interfaces, as illustrated in Fig. 5.10. It follows that taller vehicles have a performance advantage over shorter ones due to geometry.

Furthermore, the vehicle's global torsional stiffness will be dictated by the zones with the smallest local J property. Zones with a low local J property can be considered areas of low local torsional stiffness. This explains why the GSTS for convertibles is significantly lower than their roofed counterparts.

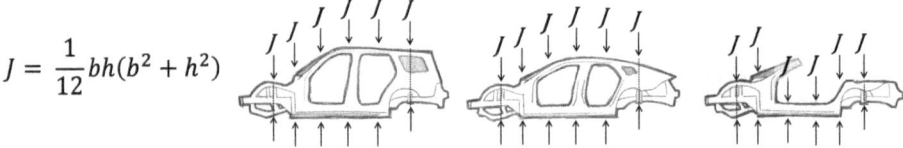

$$J = \frac{1}{12}bh(b^2 + h^2)$$

Fig. 5.10 Polar moment of inertia along the length of the vehicle

5.3.2   Application of GSTS

Comparing GSTS between automobiles of different sizes

Because the vehicle parameters of length and width are in the torsional stiffness equation, the numerical value for GSTS must be normalized to compare performance between vehicles of different sizes. To understand this, consider a 2018 Rolls-Royce Phantom VIII and a 2013 Honda Fit, as shown in Fig. 5.11.

Assuming both vehicles have an arbitrary GSTS of 30 kN m/degree, a 20 KN load would produce an angular twist based on Eq. 5.5.

$$\phi = \frac{F\left(\frac{w}{2}\right)L}{JG}\left(\frac{180}{\pi}\right) \quad \text{(degrees)} \tag{5.5}$$

The twist magnitude for the Honda would be 2.4e6/JG while the Rolls-Royce would experience 4.4e6/JG degrees of twist. Assuming similar J and G parameters for the vehicles, the Rolls-Royce would experience 83% more twist than the Honda, simply due to its larger size.

Considering that occupants perceive displacement rather than stiffness, and some aspects of noise and vibration and ride and handling performance are also sensitive to displacement, larger vehicles tend to need a higher torsional stiffness to feel and perform the same as smaller ones; the effect of differences in polar moment of inertia and/or modulus of rigidity aside.

Furthermore, the magnitude of force applied to the vehicle is influenced by the weight of the automobile, as the inertial effect of the vehicle's mass resists the force applied by the road and suspension spring, as illustrated in Fig. 5.12.

We can therefore develop an equation to normalize GSTS; Eq. 5.6. In this equation, the vehicle parameter that positively affects GSTS is in the numerator while those that negatively affect are in the denominator.

$$nGSTS = 1000\frac{(GSTS)w}{mL(9.81)} \tag{5.6}$$

	w	L
Rolls-Royce	2.0m	3.8m
Honda	1.7m	2.5m

Fig. 5.11 Vehicle size and GSTS

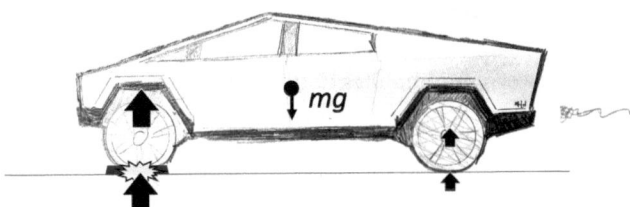

Fig. 5.12 Participation of vehicle mass

where:

GSTS	=	Global Static Torsional Stiffness (KN m/degree)
w	=	track width (m)
m	=	curb mass of the vehicle
L	=	wheelbase (m)
1000	=	conversion factor; KN to N

Equation 5.6 can be used to compare stiffness more accurately between vehicles of different sizes, as shown in Table 5.1. It can also be useful in the development of a GSTS target for an automobile under development, answering the question, 'What should the GSTS target be if we want the car to feel like its slightly smaller, lighter predecessor?'. Equation 5.7 shows how a reference nGSTS can be used to back-calculate a GSTS target.

$$GSTS\_target = \frac{(nGSTS_{reference})mL(9.81)}{w(1000)} \text{(KN m/degrees)} \qquad (5.7)$$

where:

nGSTS_reference	=	reference automobile's normalized GSTS performance
m	=	curb weight of the vehicle under development (kg)
L	=	wheelbase of the vehicle under development (m)
w	=	track width of the vehicle under development (m)
1000	=	conversion factor; KN to N

Table 5.1 Example of normalizing torsional stiffness

	GST (arbitrary) (kN m/degree)	w (mm)	m (kg)	L (mm)	nGSTS (kN m/degree)
Rolls-Royce	30	(1687 + 1671)/2	2610	3772	0.52
Honda	30	(1491 + 1476)/2	1187	2499	1.53

It should be noted that the normalized GSTS equation provides an approximated comparison between different vehicles, with regards to 'torsional feel'. The equation normalizes performance for vehicle parameters, but it does not account for differences in suspension tuning, suspension geometry, tires, or other attributes that influence how forces are modulated as they travel from the tire patch (area of tire that is in contact with the road surface) to the spring interface; all of these elements also influence displacement and an occupant's perception of torsional stiffness.

5.3.3 Structural Topology

The structural loadpath topology for the GSTS loadcase is especially difficult to parse, as so much of the structure participations, as illustrated in Fig. 5.13. Of particular importance are longitudinal shear surfaces such as the roof, floor, body-side, windshield, and sometimes even aero-panels (underside panels whose primary function is aerodynamic performance). Diagonal members that brace shear surfaces are also particularly efficient.

5.3.4 Peculiarities of BEVs

Propulsion Battery Housing Stiffness

The propulsion battery housing in the full-width underfloor configuration typically boosts GSTS performance. The amount that the PBH increases performance is dependent on how stiff the propulsion battery assembly is, the location and quantity of attachements to the body structure, and the stiffness at those body locations. BEVs with this configuration essentially have a larger local torsional stiffness in the center-compartment. GSTS increases in the range of 40% are typical of BEVs introduced in the early 2020s.

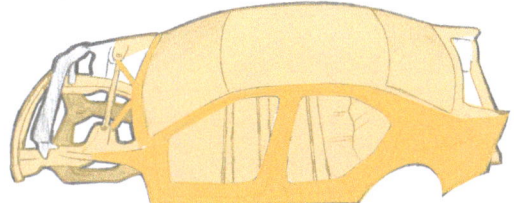 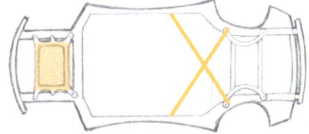

Fig. 5.13 Typical GSTS structural loadpath topology

Mass

As BEVs in the era of 600 Wh/l are heavier than their ICE counterparts, they require a higher GSTS to have an equivalent torsional feel per Eq. 5.7. Fortunately, introduction of the propulsion battery housing stiffness typically provides a GSTS boost larger than that needed due to the mass increase.

5.4   Dynamic Bending Stiffness

Just like anything else, the vehicle structure has natural frequencies which are dictated by its stiffness and mass characteristics. Manufacturers set targets on the vehicle structure's frequencies such that its vibration is consistent with the source/path/receiver Noise and Vibration tuning described in Sect. 5.2 and its vibration is not perceived by the occupant; typically, the target is to have the structure's first frequency in the range of 25–30 Hz. The first bending frequency is often of interest during the vehicle development process.

5.4.1   Loadcase Modeling

Insight from Governing Equations

As a basis, recall the fundamental natural frequency FBD and equation shown on the left side of Fig. 5.14. The natural frequency found by this equation describes the frequency at which the rigid body vibrates on the spring; a function of the object's mass and the spring rate. The natural frequency of interest in automotive design is the frequency at which the flexible vehicle structure has a resonance. Here, the mass of all hard-mounted components is considered and the boundary condition is "free-free", as if the object were floating in a weightless environment.

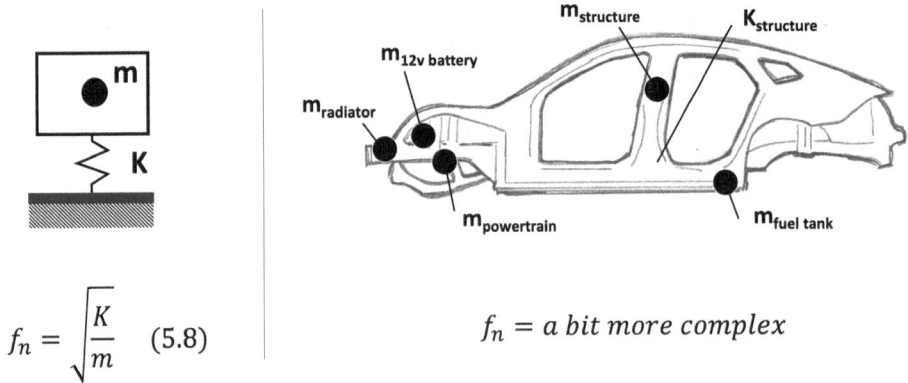

$$f_n = \sqrt{\frac{K}{m}} \quad (5.8)$$

$$f_n = a\ bit\ more\ complex$$

Fig. 5.14 Dynamic bending stiffness FDBs

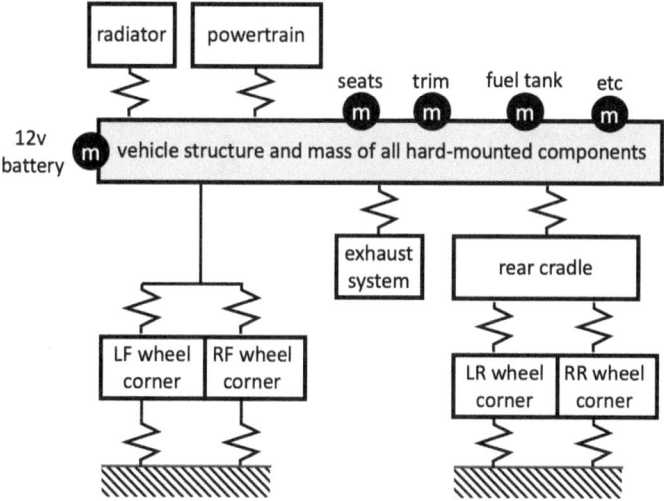

Fig. 5.15 Lumped mass/spring model of an automobile

The natural frequency in both situations is a function the mass and spring rate however determining the natural frequency for the vehicle structure is more complex due to flexibility and nonuniform geometry and mass distribution.

To understand the body of interest, consider the lumped mass/spring model of an automobile illustrated in Fig. 5.15. Vibration of the automobile is comprised of a system of sprung masses and we are interested in the highlighted mass body; the vehicle structure and mass of all hard-mounted components.

Although other natural frequencies and their associated vibration mode shapes (the deflection shape of a body undergoing a natural frequency) exist for the vehicle structure, further discussion in this chapter will focus on bending.

The simplest equation to represent the bending frequency of vehicle structure assumes that the nodes of the mode will occur at the wheel locations and is illustrated in Fig. 5.16 and written...

$$f_n = \frac{(\pi)}{2} \sqrt{\frac{EI}{qL^4}} \quad q = cw_{cw} + w_{PH-asm} \qquad [\text{Hz}] \qquad (5.9)$$

where:

EI	=	the bulk bending stiffness of the structure*
L	=	wheelbase of the vehicle
c	=	constant to adjust the participation of the curb mass
w_c	=	vehicle curb weight
w_{PB-asm}	=	weight of the propulsion battery assembly

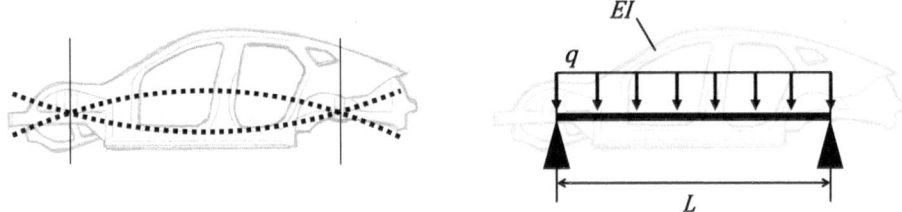

Fig. 5.16 Representing dynamic bending stiffness

$$I = \frac{1}{12}bh^3$$

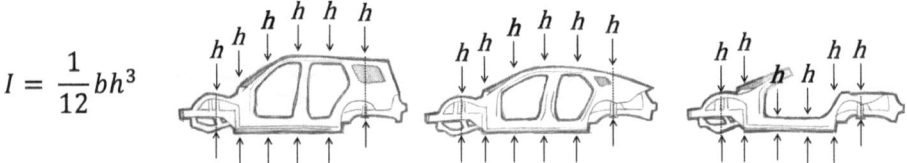

Fig. 5.17 Effect of the height of the structure

a product of material elastic modulus and the section moment of inertia

Equation 5.9 has a moment of inertia component, which tells us that taller vehicle structures will have an easier time achieving target. We can visualize the impact of height by considering the vehicle structure's moment of inertia as a compilation of discrete moment of inertia measures, as illustrated in Fig. 5.17. Performance is influenced by the weakest segment of the compilation, thus it should be no surprise that convertibles have the most difficulty achieving a high bending frequency and their targets are typically well below 25 Hz.

The equation also tells us that the bending frequency is sensitive to both the mass and the wheelbase of the vehicle structure.

As the mass increases, the natural frequency decreases. As the wheelbase of the vehicle increases, the natural frequency decreases. -this relationship will be familiar to any player of a string instrument, as illustrated in Fig. 5.18. Furthermore, it is of interest that the wheelbase is particularly influential, as it contributes exponentially in Eq. 5.9.

5.4.2 Application of Dynamic Bending Stiffness

Leveraging this equation early in the vehicle development process can be advantageous in cases where the vehicle being developed is particularly heavy and/or long. Here, a predictive equation would project the need for additional structural stiffness well before computer simulation models are available.

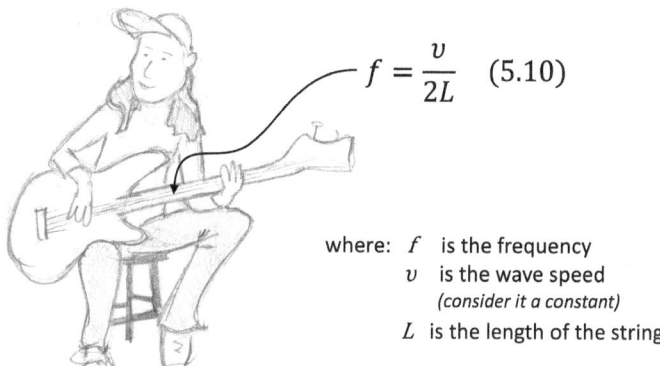

$$f = \frac{\upsilon}{2L} \quad (5.10)$$

where: f is the frequency
υ is the wave speed
(consider it a constant)
L is the length of the string

Fig. 5.18 Frequency of a guitar string analogy

Application of the equation requires development of a generalized EI. This can be achieved by back-calculating EI values using parameters and performance of known vehicles (reference vehicles). Ideally, a different generalized EI value would be developed for different vehicle types; sedans, SUVs, etc.

Table 5.2 illustrates the process where an generalized EI value is developed and applied to a vehicle under development. The first two rows show EI value calculation for two reference vehicles, using their known vehicle parameters and bending frequency. The third row shows a bending frequency estimation calculated for a future BEV sedan using an average of the two reference vehicle EI values. In this case, performance projection for the 'future BEV Sedan' suggest that additional structure will be required to meet a frequency target of 25 Hz or greater.

Table 5.2 Projecting dynamic bending stiffness of a vehicle under development

Vehicle	f_n		EI		q	Vehicle curb weight	PBH-asm weight	L
	Actual	Predicted	Calculated	Generalized				
BEV sedan A	28	–	11,602,900	–	502	2604	684	2092
BEV sedan B	30	–	8,999,000	–	422	2217	502	2077
Future BEV sedan	–	24	–	10,200,000	533	2870	700	3.00

Fig. 5.19 Typical dynamic
stiffness structural loadpath
topology

Dynamic Bending Stiffness versus Vehicle Mass
(EI = 11602900, L = 2.92m)

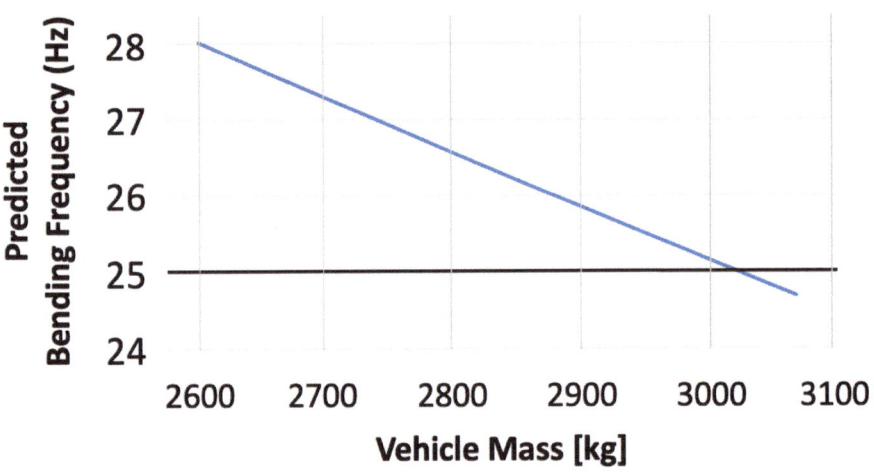

Fig. 5.20   Sensitivity of performance to vehicle mass

5.4.3   Structural Topology

The structural loadpath topology for dynamic bending is rather simple to visualize; it consists of any longitudinal member that can contribute to global bending stiffness or vertical members that create a vertical truss system, as illustrated in Fig. 5.19.

5.4.4   Peculiarities of BEVs

Mass
Many ICE vehicle structures meet their targeted bending frequency without the need for specific stiffening elements, as development for other loadcases result in a sufficiently stiff structure. BEVs, however, are heavier than their ICE counterparts in the era 600 Wh/l cell

chemistry and meeting a bending frequency target can become a loadcase which drives the global structure design, simply due to its increased mass, as Fig. 5.20 illustrates.

5.5   Roof Crush

The Roof Crush Resistance loadcase exists as both regulatory and consumer metric tests. FMVSS216a is an example of the loadcase being regulatory and the IIHS Roof Strength test is an example of it being a consumer metric measure. The IIHS introduced their test after studying rollover accidents and finding a relationship between fatalities and the strength of the roof. Their performance threshold was more stringent than existing tests and remains the most structural demanding at the time of this book's publication. The loadcase is often referred to as "Roof Crush" or "Roof Strength".

Although there are significant differences between test protocols, there are also commonalities. In all protocols, the test is conducted using full vehicle which is supported at its *rockers*. A surface, or platen, translates along a prescribed axis at a relatively low speed and contacts the test vehicle. The platen continues to translate towards the vehicle until it reaches prescribed distance, after which the platen retreats to its original position. The loadcase condition is illustrated in Fig. 5.21. Link 5.1 provides an overview in video format.

The platen is angled in the test fixture in front and side views, as illustrated in Fig. 5.22.

Fig. 5.21   General roof crush
loadcase condition

Fig. 5.22 Platen angles

	Roof-Crush of a 2022 Volkswagen GTI (http://www. youtube.com/watch?v=t0InhxbyJdg) IIHS YouTube channel	(Link-5.1)
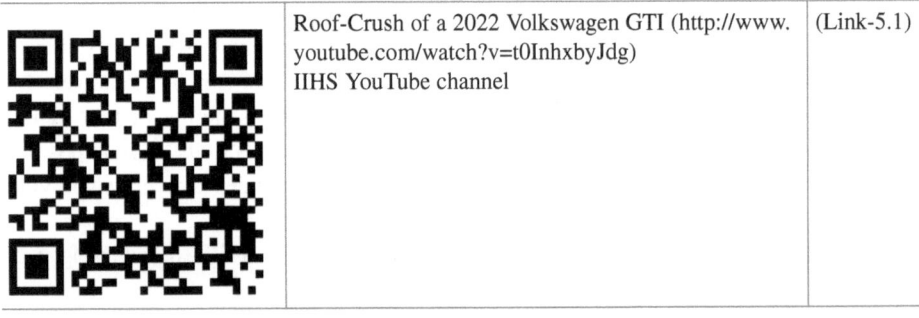		

The test fixture and platen are instrumented such that reaction force is measured as the platen translates, and force versus platen displacement data is produced, as illustrated in Fig. 5.23. Results are typically shown in the form of 'strength to weight ratio' (SWR) versus displacement. Here, SWR is simply the reaction force divided by the vehicle weight (mg), as shown in Eq. 5.11.

$$SWR(\delta) = \frac{F(\delta)}{mg} \qquad (5.11)$$

where:

F = reaction force seen by the test fixture

δ = platen displacement

m = vehicle mass

g = acceleration of gravity

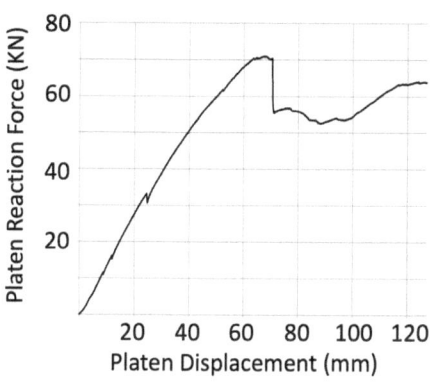 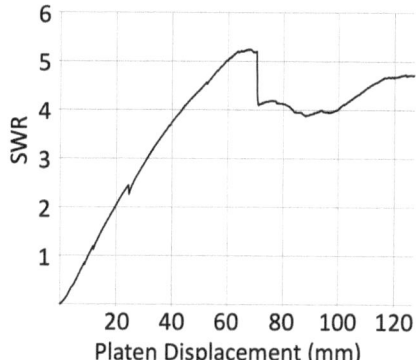

Fig. 5.23 Example Roof Crush test results

In practice, the term SWR is used to signify the maximum value of SWR(δ); a single value. The performance of the vehicle described in Fig. 5.2 would be stated as, "5.2 SWR".

The rating system for each protocol requires that a threshold SWR value be achieved within a specified platen displacement. The FMVSS16a requires that vehicles weighing less than 2722 kg (6000 lb) achieve an SWR of 3.0 or higher within 127 mm (5 in) of platen travel while the IIHS Roof Strength test has a graduated rating system based on a vehicle's SWR within 127 mm of travel, as shown in Table 5.3—note that tests conducted by the IIHS crush the test vehicle to 254 mm (10 in) but ratings are based only on data from the first 127 mm of crush.

The vehicle curb weight (no occupant, no cargo, full fluids) is used for the SWR calculation, however test protocols do not specify specific vehicle option content. Hence, an automobile could be tested in its AWD or its 2WD form, with its 4-cylinder engine or its 6-cylinder engine option, etc. It is important to understand that the SWR value reported for any test or assessment is dependent on the vehicle weight used in the calculation. It is typical for manufacturers to engineer structural performance assuming the heaviest possible curb weight for an automobile model to ensure performance for all option configurations of that model. The heaviest possible curb weight is often called 'max curb weight'.

Table 5.3 IIHS rating protocol (version V)

SWR	IIHS roof strength rating
$4.00 \leq$ SWR	Good
$3.25 \leq$ SWR < 4.00	Acceptable
$2.50 \leq$ SWR < 3.25	Marginal
SWR < 2.50	Poor

The Roof Crush loadcase is performed in a "one-sided" or "two-sided" procedure, depending on protocol. The IIHS Roof Strength test, for example, is a one-sided test, where the vehicle is placed in the fixture and only one side of the vehicle is crushed. The FMVSS216a is a two-sided test where one side of the vehicle is crushed until the required reaction force is achieved. The opposite side of the damaged vehicle is then crushed to 127 mm of platen displacement (Link 5.2).

FMVSS216a is also an example of a test procedure that considers headroom. In this test, the required SWR must be met before 222N (50 lb) of force is seen by a headform representing the 50% percentile male position.

Although Roof Crush is performed on a complete vehicle it is technically a subsystem loadcase. The test is performed at a vehicle level because it would be costly to de-content the vehicle and the strength and stiffness associated with the nonstructural parts are insignificant to the performance.

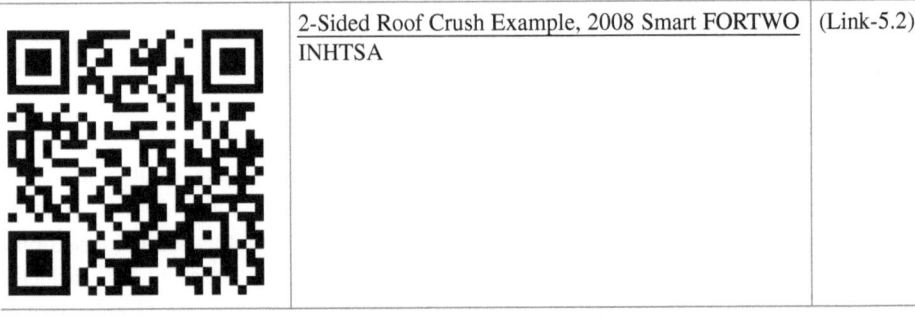

| | 2-Sided Roof Crush Example, 2008 Smart FORTWO | (Link-5.2) |
| | INHTSA | |

5.5.1 Loadcase Modeling

A model and FBD of the roof crush test are illustrated in Fig. 5.24.

- Plasticity (s): Plasticity exists in the test vehicle and crush is inevitable due to the definition of the test protocol.
- Friction (f): Friction is present between the platen and the test vehicle.
- Elasticity (K_X, K_Y, K_Z): The test vehicle has stiffness which can be separated into vertical and lateral components, K_Z and K_Y respectively. The fore-aft component of platen force might suggest consideration of a longitudinal stiffness K_X, however we will find that this is rarely beneficial.
- Boundary Condition: The test vehicle is constrained at the rockers

Fig. 5.24 Roof crush model/ FBD

5.5.2   Fundamental Behavior and Physics

Performance in the Roof Crush is measured by the force which the test vehicle resists the platen displacement. Recall the sponge experiment of Sect. 3.2.3.2 where the stiffness of the loadpath dictates the reaction force. The goal of vehicle structure in this loadcase is to be stiff enough to reach the desired reaction force magnitude. Also recall the paper cup experiment of Sect. 3.2.3.3, illustrating that strength of the loadpath can also dictate the reaction force. Thus, the goal is also to delay plastic deformation within the loadpath until the desired stiffness is achieved.

Event Milestones

The event milestones of Roof Crush are illustrated in Fig. 5.25.

Behavior, High-Level

(a) The displacing platen contacts the test vehicle and the test begins
(b) Minor yielding within the structure creates momentary reductions of reaction force. If the slope after the yielding is equivalent to before, the yielding is typically irrelevant; the collapse of the thin *body-side-outer-panel* onto the *A-pillar-reinforcement*, for example.
(c) Two high-level deformation modes occur as the vehicle deflects; axial compression and bending of the B-pillar, and match-boxing of the upper-structure. The stiffness of the structure reduces as the structure matchboxes.
(d) Windshield fracture
(e) Buckling of structural elements within the loadpath; *B-pillar* bucking, for example.

Behavior, Detailed

Vehicle behavior in Roof Crush is often characterized by the match-boxing of its upper structure and the axial compression and bending of the *B-pillar*. Match-boxing occurs due to the lateral component of the platen force, while *B-pillar* compression occurs due to the vertical component. As the test progresses, the platen pushes the *roof-rail* inboard and down. The inboard displacement results in the *B-pillar* load condition becoming less axial and more bending, as illustrated in Fig. 5.25c. Increased *B-pillar* bending and upper structure match-boxing, both fundamentally softer conditions than axially loading the *B-pillar*, result in the softening of the structure as the test progresses. A reduction of the

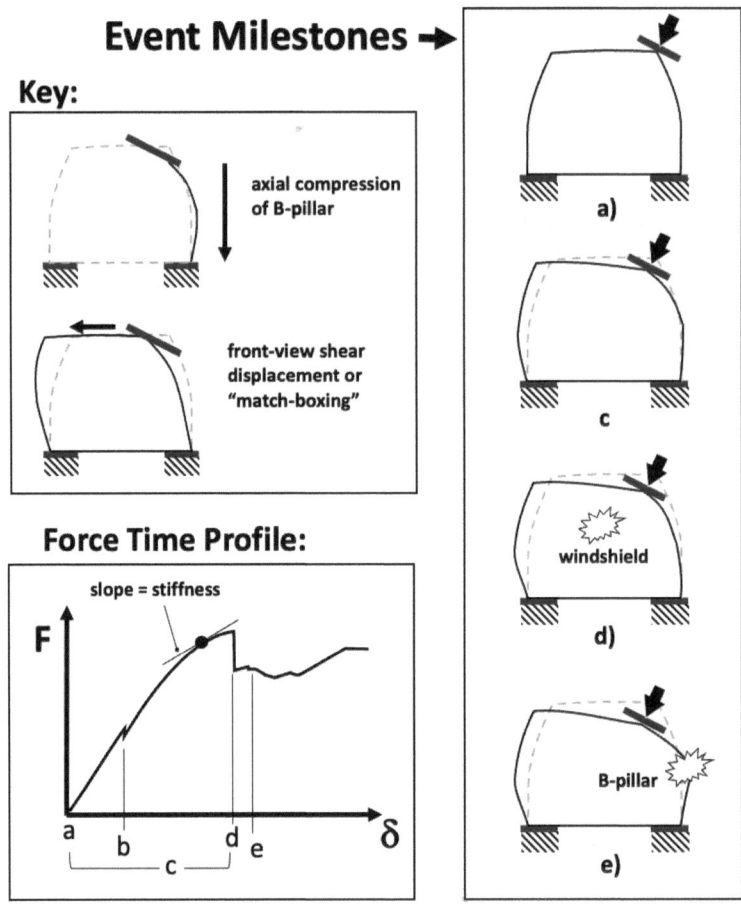

Fig. 5.25 Roof crush event milestones

F(δ) response curve slope indicates such softening, as seen in Fig. 5.25 just prior to point d.

Vehicles with a more inclined side structure and *B-pillar*, more "tumblehome" (as illustrated in Fig. 5.26), are disadvantaged with regards to the balance between *B-pillar* axial compression and upper structure match-boxing. Stiffer and stronger local structure can be required for vehicles with a large amount of tumblehome, as the key to efficient roof strength performance is to keep the *B-pillar* upright and delay section buckling and windshield fracture.

At some point in nearly every test, the windshield fractures and a drop in reaction force and stiffness accompanies this event. Local yielding or section buckling of the *front-header* or *A-pillar* often occurs at the same time as windshield fracture and it can often be difficult to determine which instigated the failure, as the windshield and the surrounding

 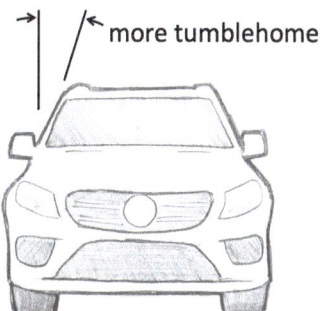

Fig. 5.26 Tumblehome illustrated

Fig. 5.27 Side view shear

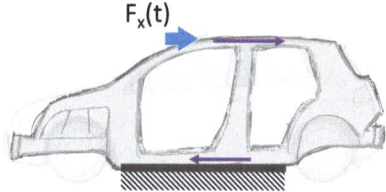

body structure act jointly to provide shear strength; if one fails, the other cannot manage the force alone.

The windshield and surrounding body structure also provide a significant shear stiffness loadpath to resist match-boxing of the upper structure. It is for this reason that the vehicle stiffness after windshield fracture is not as stiff as before, and the peak SWR value typically occurs prior to windshield fracture.

Shear is also occurring in the longitudinal direction due to the fore-aft component of the platen force, as illustrated in Fig. 5.27. Considering that the force magnitude in this direction is relatively low and that a conventional body structure is relatively stiff in a side view shear direction, consideration of this force component can be typically ignored.

Detailed behavior of a vehicle structure in the Roof Crush loadcase can vary significantly depending on the vehicle's geometry and construction. During computer simulation, it is often helpful to monitor section forces within the loadpath and examine how they change as the procedure progresses.

5.5.3   Structural Topology

The structural loadpath topology for Roof Crush is predominantly the *B-pillar*, the *A-pillar* and *roof-rail*, the windshield and the *front-header*, as illustrated in Fig. 5.28.

Fig. 5.28 Typical roof crush structural loadpath topology

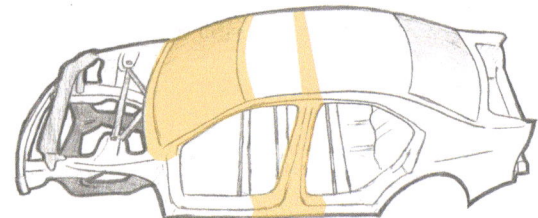

Roof-Rail

The function of the *roof-rail* is to the transmit platen force to the front-header/windshield and the B-pillar. An efficient *roof-rail* is stiff, to maximize the amount of transferred force, and strong, to transfer that force as long as possible. Effective *roof-rail* sections tend to be deep in the direction of platen travel and do not have geometry that would instigate section buckling, as illustrated in Fig. 5.29. Typically, it is more important for the section to have high buckling resistance (Fig. 5.29b) than to maximize the I-value and stiffness (Fig. 5.29a).

Windshield

One of the goals of efficient design is to delay windshield fracture as long as possible. Thus, buckling resistance of the *front-header* is essential. Efficient *front-header* designs will be as deep as possible in the vertical direction, avoid geometric features that would instigate buckling, and include geometric features to boost buckling resistance. Bending strength of the *A-pillar/roof-rail* to the *front-header* joint is also critical, however it can be tricky to engineer considering manufacturing constraints. In nearly all vehicle designs, windshield fracture is inevitable, as the platen will eventually contact and apply direct loading to the windshield.

B-Pillar

Design elements for *B-pillar* stiffness and strength in the axial and bending directions vary depending on the general *B-pillar* shape as dictated by the side-door, seat belt system, and

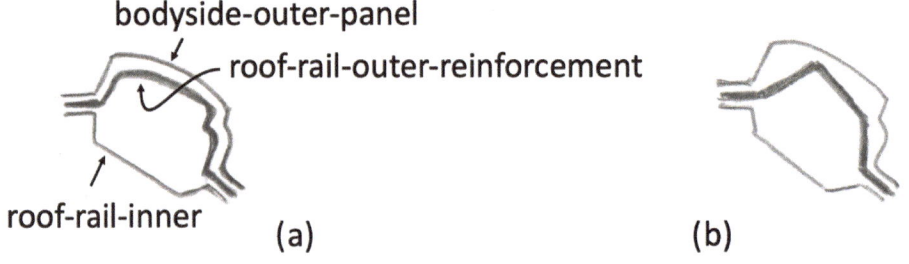

Fig. 5.29 Roof rail section design elements

tumblehome. Although buckling tendencies can vary between different *B-pillar* shapes, three design elements and principles typically apply to all:

(a) The *B-pillar-inner* is on the compression side of the section as is undergoes axial compression and bending. Maximize its strength. This can be done with efficient geometry; avoid strength discontinuities, incorporate strengthening geometric features, material properties, etc. Note the continuous, vertical geometric trough feature in Fig. 5.30. The *B-pillar-inner*, however, is on the tension side of the section for the Side MDB loadcase, and a good *B-pillar-inner* design balances the needs of both loadcases.

(b) *B-pillar* buckling often occurs at the beltline height (Fig. 1.1). The mode of buckling is such that the section becomes unstable and the opposite flanges move away from each other, flattening the section, as illustrated in Fig. 5.31a. An efficient design element to prolong section stability is a *tension-panel* which resists this mode shape, as illustrated in Fig. 5.31b.

(c) visualize the *B-pillar* as a bridge. Focus on creating planar shear surfaces, slowly transitioning surfaces, and avoiding other strength discontinuities. Engineer gradual changes in strength along the B-pillar's height.

You might have noticed that the design elements have been described in the order in which they typically participate in the Roof Crush loadcase. There is a tendency to look at the deformed shape of a vehicle that has undergone Roof Crush testing and attempt to address each of the observed structural failures. This strategy is will not lead to an efficient design

Fig. 5.30 B-Pillar-inner design elements

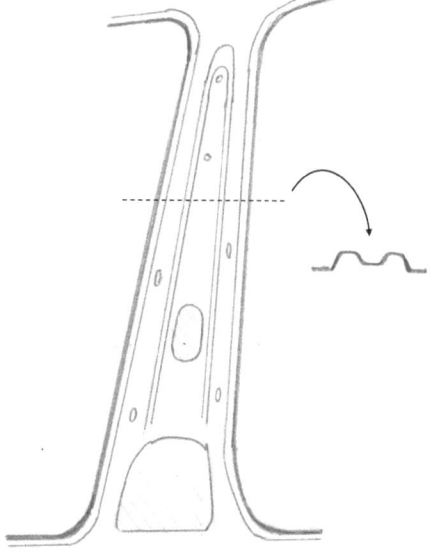

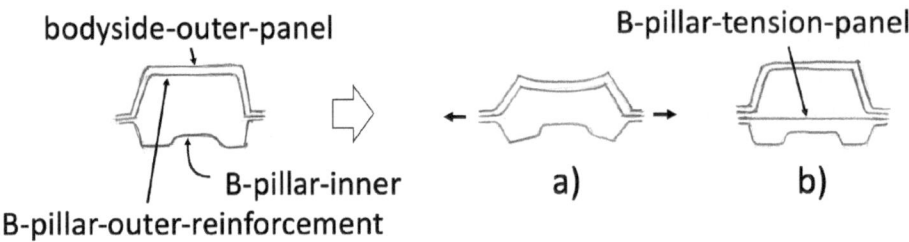

Fig. 5.31 B-pillar buckling and tension-panel design element

for two reasons; Roof Crush is a prescribed displacement test and structural failure is therefore inevitable, and the order in which the structural failures occur is very important. It is best to understand the structural function as...

• Provide a stiff and strong *roof-rail* section to maximize load transfer and maintain it as long as possible.
• Provide a stiff and strong *front-header* section to delay windshield fracture for as long as possible.
• Incorporate *roof-rail* to *roof-bow* design elements that maximize match-boxing stiffness.
• Provide a stiff and strong *B-pillar* maximize load transfer and maintain it as long as possible.

...and focus performance development on the first structural failure that occurs in physical test or computer simulation. This strategy requires a recursive execution of performance development, but it ensures that only the issue limiting performance is being addressed.

5.5.4 Peculiarities of BEVs

Mass
BEVs have no inherent modification to the Roof Crush structural topology, however BEVs in the era of 600 Wh/l cell technology are heavier than their ICE counterparts and this translates to higher SWR requirements and consumer metric targets.

Unique EV Loadcases

6

Abstract

Previous chapters of this book have addressed how the EV configuration impacts individual loadcases and how the automotive structure might adjust accordingly in each case. The introduction of high voltage content into the automobile has required, however, some additional consideration and introduction of EV-specific loadcases. This short chapter provides an overview of such loadcases which have implications on structural strategy and design.

Two fundamental dangers of High Voltage content are fire resulting from short circuits and the fact that direct contact with high voltage can be deadly to humans. The FMVSS305 and UNECE regulations are in place to ensure post-crash high voltage safety in specific loadcases and manufacturers consider high voltage safety in the other forementioned loadcases. However additional loadcases are typically considered by a manufacturer so as to provide a more comprehensive safety package. These additional loadcases might include the following conditions.

Underside protection

There are many types of situations where the underside of an automobile comes into contact with something and these situations can be concerning for EVs with a propulsion battery packaged underfloor. Although these situations tend to occur rarely, it is important to consider protection of an underfloor propulsion battery. Underside protection loadcases

© The Author(s), under exclusive license to Springer Nature Switzerland AG 2025
M. Dingman, *Fundamentals of Automotive Structures and Battery Electric Vehicle Applications*, Synthesis Lectures on Mechanical Engineering,
https://doi.org/10.1007/978-3-031-75933-8_6

Fig. 6.1 Full width load condition examples

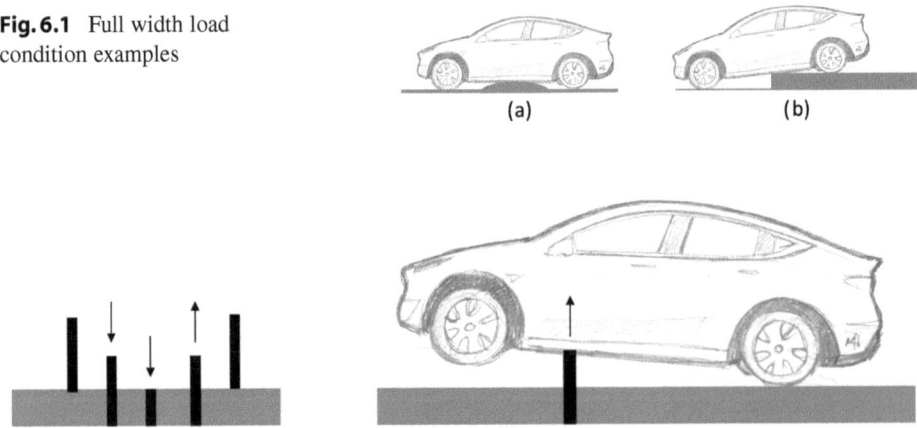

(a)   (b)

Fig. 6.2 Bollard action and a deployed bollard condition

are not standardized or regulated at the time of this book's writing, however manufacturers develop and consider loadcases that fall into two general categories; full-width loading and localized loading.

Full Width Loading

Loadcase events in which the entire width of the vehicle underside is contacted by something fall into the category of "full-width loading". Examples of such conditions are speed bumps and 'end of road' conditions, as illustrated in Fig. 6.1a and b respectfully.

Localized Loading

Conditions where only a small area of the vehicle underside is contacted are classified as "localized loading". Situations where an automobile drives over road debris is an example. The road debris can bounce vertically into the vehicle underside and can be of any imaginable shape, strength, and density; the possibilities are endless.

A manufacturer can expect more frequent debris-type loading if the automobile is intended to be used off-road. Rock contact loadcases have been developed and assessed for such vehicles.

The 'Bollard Test' represents another type of localized loading. Bollards, the strategically distanced poles used to restrict automobile access, are sometimes retractable. The condition where an automobile is positioned over a retracted bollard when it is deployed, as illustrated in Fig. 6.2, is one that has been considered when developing BEVs.

Manufacturers have also shown consideration for the localized loadcase where a customer attempts to lift their vehicle with a floor jack positioned under the battery.

Due to the infrequent nature and variability of these events, loadcase development can require a fair amount of judgement and be quite a challenge.

Underside protection loadcases are an important consideration during structural development, as they can influence how the propulsion battery housing is packaged relative to the body structure and the detail design of the battery housing structure itself.

Thermal Runaway Protection (TRP)

TRP represents the scenario where a single battery cell is shorted and the resulting fire has the ability to ignite neighboring cells (runaway). The goal of the battery housing structure is to prevent fire from entering the occupant compartment for an adequate amount of time. TRP is not a loadcase that has significant influence on the structural loadpath topology or strategy, however it can influence detailed material execution of the propulsion battery housing.